Gramsci is Dead

Gramsci is Dead

Anarchist Currents in
the Newest Social Movements

Richard J.F. Day

Pluto Press
LONDON • ANN ARBOR, MI
and
Between the Lines
TORONTO

First published 2005 by Pluto Press
345 Archway Road, London N6 5AA
and 839 Greene Street, Ann Arbor, MI 48106
www.plutobooks.com

and
Between the Lines
720 Bathurst Street, Suite 404, Toronto, Ontario M5S 2R4
www.btlbooks.com

British Library Cataloguing in Publication Data
A catalogue record for this book is available from the British Library

ISBN   0 7453 2113 5 hardback
ISBN   0 7453 2112 7 paperback (Pluto Press)
ISBN   1 897071 03 5 paperback (Between the Lines)

Library of Congress Cataloging-in-Publication Data
Day, Richard J. F.
   Gramsci is dead : anarchist currents in the newest social movements /
Richard J.F. Day.
      p. cm.
   Includes bibliographical references and index.
   ISBN 0–7453–2113–5 — ISBN 0–7453–2112–7 (pbk.)
   1. Radicalism. 2. Socialism. 3. Communism. 4. Anarchism. 5. Poststructuralism. I.
Title.
   HN49.R33D39 2005
   322.4—dc22

                                                                    2005001486

Library and Archives Canada Cataloguing in Publication
Day, Richard J.F.
Gramsci is dead / by Richard Day.
Includes bibliographical references.
ISBN 1–897071–03–5
1. Anti-globalization movement. 2. Philosophy, Marxist. 3. Anarchism.
4. Poststructuralism. 5. Social movements—Political aspects. I. Title.

HN49.R33D395 2005   322.4   C2005–902450–X

Between the Lines gratefully acknowledges support for our publishing programme
from the Canada Council for the Arts, the Ontario Arts Council, the Government of
Ontario through the Ontario Book Publishers Tax Credit program and through the
Ontario Book Initiative, and the Government of Canada.

10 9 8 7 6 5 4 3 2 1

Designed and produced for Pluto Press by
Chase Publishing Services Ltd, Sidmouth, EX10 9QG, England
Typeset from disk by Stanford DTP Services, Northampton, England
Printed and bound in Canada by Transcontinental Printing

Contents

Acknowledgements

I could not have written this book without drawing upon the labours of student/activist researchers who have been associated with the Affinity Project, which operates out of Queen's University at Kingston, Ontario and maintains a website at <www.affinityproject. org>. These in include Hilton Bertalan, Naila Bhanji, Jake Burkowicz, Enda Brophy, Christopher Canning, Ryan Mitchell, Rick Palidwor, Nadia Knircha, Brianne Selman, Andrew Stevens, and Lori Waller. I have also benefited from conversations with Taiaiake Alfred, Allan Antliff, Mark Coté, Glen Coulthard, Greig dePeuter, David Firman, Dina Khorasanee, Jennifer Pybus, and Scott Uzelman, and from interactions with undergraduate and graduate students in my seminars at Queen's University. I am particularly indebted to Sean Haberle, who traveled extensively to conduct interviews with activists at various sites around central Canada and the northeastern United States. Sarita Srivastava also deserves special mention for suggesting a major revision that got me past a particularly daunting writing obstacle, and for being an inspiring colleague and friend. Finally, and most importantly, I want to acknowledge the contributions of my partner Alison Gowan, who has for many years been a crucial source of intellectual and emotional support for my work, and much more importantly, for my life as a whole.

Introduction

Seek and learn to recognize who and what, in the midst
of inferno, are not inferno, then make them endure, give
them space.

Italo Calvino, *Invisible Cities*

SEATTLE, ANARCHISM AND THE CORPORATE MASS MEDIA

For most people in the G8 countries, the Seattle anti-World Trade
Organization (WTO) protests in late 1999 mark the point at which a
new militancy erupted onto the surface of an otherwise serene liberal-
democratic polity. I was living in Vancouver, British Columbia at the
time, and decided at the last moment not to go to Seattle because I
had a huge pile of papers to grade. Just another protest I thought, and
I'd have to sit on a bus for five hours to get there and back. At one
point in the day, though, I turned on the television just to see what
was happening. I was fascinated and surprised by the now famous
images of huge marches, lockdowns, roving bands of riot police,
with endless clouds of tear gas shrouding the scene. Over this scene
from another (part of the) world, came the voice of a local reporter
out in the streets:

> *Reporter:* 'There are some people here, roaming about … well not exactly
> roaming, they seem organized. I don't know who they are, they're all
> dressed in black, they have black hoods on, and black flags … a flag with
> nothing on it.'
> *Anchor:* 'A flag with nothing on it?'
> *Reporter:* 'That's right, it's totally black.'

Thus did the Black Bloc make its first appearance on the North
American mass-media stage. After a while, the TV stations in Seattle
thought they had it figured out: these were 'anarchists', whatever *that*
might mean. No one tried to confirm this with the masked protesters
themselves, however—they looked far too dangerous, and despite the
fact that they were at a mass protest, it seemed they did not want
anyone to know who they were.

As more reporters—as well as political commentators, police chiefs
and academics—began to weigh in on what happened in Seattle,

the empty expanse of that black flag was filled in with the help of stereotypes, prejudices and fragments of knowledge from the previous century. A *Time* magazine story of December 13, 1999 contained a section on 'The Violence', with the subhead 'How Organized Anarchists Led Seattle into Chaos'. After dismissing the protesters as 'a sprawling welter of thousands of mostly young activists populating hundreds of mostly tiny splinter groups espousing dozens of mostly socialist critiques of the capitalist machine', the article tantalized and appalled its readers with images of circle-A placards, store looting and an unidentified person, with no insignia of any kind, climbing through a broken window. The caption read 'Coffee to go: An anarchist patronizes a Seattle Starbucks'.

Since the Battle of Seattle, this one-sided, ill-informed caricature of anarchist activism has become almost obligatory in the corporate mass media. The *Vancouver Sun*, a right-wing daily, ran a full-page article on the protests against the Organization of American States in Windsor, Ontario in its issue of June 3, 2000 (p. A14). It appeared under the title 'For the New Anarchists, the Message is the Mayhem' and included the subheads 'Organized Radicals' and 'One Man's Philosophy'. It describes anarchism as a 'long-dormant philosophy' that has returned with 'destructive force' and 'disruptive power'; words like 'threat', 'mayhem' and 'rampage' pepper the article. It also attempts to belittle the well-known Canadian activist Jaggi Singh by noting that, before he sits down to talk with reporters, he orders a coffee from a Canadian doughnut-shop chain: 'It turns out that even anti-corporate anarchists appreciate their Tim Horton's'. A similar line was taken by the Vancouver *Province* (the only other daily newspaper in Vancouver, which is owned by the same corporation as the *Sun*) in an article after about the inquiry into police brutality at the 1997 APEC Summit in that city.[1] The headline read: 'APEC Protest Leader Stays in Hotel Van[couver], like Suharto' (May 3, 1999: A3). This is an extremely common trope of exclusion by inclusion, which works by trying to show that They (anarchist activists) are no less tainted with the stain of capitalist individualism than We (good capitalist citizens) are, and therefore have no right to criticize the status quo.

Of course, fighting these highly functional (mis)representations might be seen as a small problem, as the so-called 'anti-globalization movement' is now receiving much less media attention. While Seattle was followed by major protest convergences at Washington, Genoa, Prague and Quebec City, the police in the 'advanced liberal democracies' now seem to have discovered that making unconstitutional arrests and closing borders to anyone wearing a bandana are sufficient to squelch what used to be considered legitimate dissent. Indeed, the response to 9/11, coupled with the ongoing wars in Iraq and Afghanistan, has dealt what looks like a death blow to the most visible expressions of resistance to neoliberalism in the global north. This could mean that a time of mourning is at hand; but even though much has been achieved over the past few years, many do not lament the demise of this phase of struggle. Despite the fact that they have helped to raise awareness of the dark side of the new world order, summit protests are limited, when they 'work' at all, to temporarily impeding or slightly reforming existing structures. Though they may build skills and structures that prefigure alternatives, they are not capable of addressing the fundamental problems associated with the expansion and consolidation of the racist, heterosexist, system of neoliberal-capitalist nation-states. Also, as has been pointed out quite often in recent years, summit-hopping is an elitist practice that drains time and energy from local communities (Marco 2000; Pastor and LoPresti 2004: 29).

My own doubts about the efficacy of the protest convergence model were solidified in July 2003, when I arrived at the protest against the Montreal WTO mini-ministerial. Riot cops had surrounded the green zone with red tape and were arresting everyone who refused to leave the area. As I approached the line, one particularly nasty fellow, standing in riot gear with his legs spread, stared at me and whacked his club against the palm of his hand. I took out a pen and my notebook and repeated the gesture for him. This book is best thought of as an extended version of that encounter: now that the cops of the G8 countries are no longer surprised by direct action tactics, now that their political masters are willing to broadly adopt the repressive tactics of what are hypocritically called 'Third World dictatorships', how will the struggle against globalizing capital—and the many systems of domination and exploitation with which it is inextricably linked—continue? Among indigenous peoples and in the global South the answer seems clear: it will continue as it always has, for hundreds of years, taking a multiplicity of forms, including of course the kinds of mass, violent protests that so shocked G8 sensibilities when they happened in Genoa rather than Jakarta. Seen as nothing more than a violent clash between protesters and police, the only thing special about Seattle was that it happened where it did. But this point of view is a bit too simplistic. Even if the cycle of mass protest convergences in the global North has abated, the ideological and organizational structures that brought them about have hardly disappeared. It is to these deeper, broader and longer-running currents that we must turn our attention if we are to understand where 'the anti-globalization movement' has come from, what it has done, and where it might be headed.

NAMING 'THE MOVEMENT' ... AND OTHER KEY TERMS

Before embarking on this inquiry, though, a number of preliminary issues must be addressed. One of these is the question of whether the new forms of activism that are emerging to contest neoliberal hegemony are best analysed as movements against globalization as such. Many commentators and activists have argued—and I would tend to agree with them—that to proceed in this way is to fall into an analytic trap (N. Klein 2001; Buchanan 2002; Milstein 2002). How, then, are we to refer to these resurging and abating struggles, how do we discuss them without doing violence to what they stand for? What, indeed, do they stand for—and against—if they are so difficult to encapsulate in a single term? These are questions that cannot be answered simply or quickly; yet *some* provisional assumptions must be made. To this end, while I accept that the term 'anti-globalization' speaks to important concerns about capitalism, colonialism and democratic accountability, it is also clear that these concerns do not cover the entire spectrum of resistance to the new world order. Rather, I see them as representing particular sites of condensation within a much more complex field of *contemporary radical activism.*

By contemporary, I mean primarily of the late 1990s and early 2000s, but with roots reaching back to the new social movements of the 1960s—feminisms, the US civil rights movement, Red Power, anti-colonialism, gay and lesbian struggles—as well as to 'older' traditions of marxist and anarchist socialism. By radical activism I mean conscious attempts to alter, impede, destroy or construct alternatives to dominant structures, processes, practices and identities. My focus is quite literally those struggles that seek change to the root, that want to address not just the *content* of current modes of domination and exploitation, but also the *forms* that give rise to them. Thus, for example, rather than seeking pay equity for men and women, a radical feminism works for the elimination of patriarchy in all of its forms; rather than seeking self-government within a settler state, a radical indigenous politics challenges the European notion of sovereignty upon which the system is states is constructed. Contemporary radical activism, then, pushes beyond the possibilities and limits of liberal reform, while not entirely discrediting attempts to alter the status quo—one can never be sure of the value of a strategy or tactic without reference to particular social, historical and political contexts. At the same time, it does not seek a return to the theory and practice of the Old Left of the nineteenth and early twentieth centuries, or even to the New Left of the 1960s to 1980s. There is something else going on here, something different, which I try to indicate by sometimes using the term *newest social movements* to describe those currents in which I am most interested.

Understanding what these movements stand for and against is another necessary but perilous endeavour. There is much disagreement on this subject within and across activist communities, and scholars have not had any greater luck in reaching a consensus. With the widespread enthusiasm generated by Michael Hardt and Antonio Negri's *magnum opus* (2000), the term Empire is gaining in currency, within academic circles at least, as a shorthand description of the common enemy. *Empire* has the merit of analysing not only the structures and processes of globalizing capital, but also the system of states and superstates and the burgeoning societies of control that are intimately bound up with them. However, inasmuch as this rather lengthy text fails to address questions of gender in any systematic way, and speaks of racism only in a seven-page section, it has been criticized as presenting a Eurocentric, androcentric—and perhaps a class-determinist—account (Mishra 2001; Moore 2001; Quinby 2003). Although Hardt and Negri have promised to address these issues in greater detail in future work together, their concept of the multitude as a global proletariat seems very difficult to reconcile with postmarxist critiques of a politics that gives centrality to the struggles of the working class, and with anti-racist feminist calls for the decolonization of theory and the practice of solidarity across all axes of oppression (Arat-Koc 2002; Mohanty 2003). Also, the reception of *Empire* has been much less enthusiastic among activists, many of whom do not appreciate its difficult prose and its implicit reliance upon a network of unfamiliar concepts developed by Italian autonomist marxists (Flood 2002; los Ricos 2002).

For these reasons I am reluctant to deploy the term Empire in my own work. I am also reluctant to rely upon a neologism, however, since this will not solve the fundamental problem, which is that *there is no single enemy* against which the newest social movements are fighting. Rather, there is a disparate set of struggles, each of which needs to be addressed in its particularity. Yet the actuality of globalizing capital and the intensification of the societies of control mean that all these struggles occur in an increasingly common context, even if they do not explicitly identify this context, or elements of it, as what they are struggling against. Thus, while I do not wish to totalize or reduce the diversity of contemporary struggles, I will refer to *the neoliberal project* as providing a shared background or context within which they occur. The neoliberal project includes the ongoing globalization of capital, as well as the intensification of the societies of control; it also relies upon and perpetuates shifts in the organization of the system of states, through regional agreements such as the North American Free Trade Agreement (NAFTA), and the construction of superstates such as the European Union. This is to say that I take as given that we cannot understand state domination outside of capitalist exploitation, and that we cannot understand either of these phenomena without reference to the societies of control.

It should also be noted that state domination and capitalist exploitation would be impossible if it were not for the fact that neoliberal societies are divided according to multiple lines of inequality based on race, gender, sexuality, ability, age, region (both globally and within nation-states) and the domination of nature. Populations must be sorted into apparently 'natural' hierarchies if the differential distribution of social goods that capitalism creates is to be reconciled with the values espoused by a liberal politics. Because these hierarchies must be strengthened as liberalism transforms itself into neoliberalism and inequalities increase, we have seen a return of social conservatism and a backlash against the progressive change brought about during the heyday of the Keynesian welfare state. When I refer to the neoliberal project, then, I am hoping to describe a complex web of practices and institutions that have the effect of perpetuating and multiplying various forms of interlocking oppression (hooks 1984; Collins 1991). These allow 'populations' to be divided and managed, and our daily lives to be more intensely immersed in capitalist accumulation and rational-bureaucratic control (Foucault 1991).

Thinking about the neoliberal project leads directly to a consideration of the problem of *hegemony*. Although I will have much to say about the changing meanings of this term over time, for the moment it can be taken as describing a process through which various factions struggle over meaning, identity and political power. To use the words of Antonio Gramsci, a key thinker in this lineage, a social group which seeks hegemony strives to 'dominate antagonistic groups, which it tends to "liquidate," or to subjugate perhaps even by armed force', at the same time as it attempts to 'lead' kindred and allied groups (Gramsci 1971: 57). Hegemony is a simultaneously coercive and consensual struggle for *dominance*, seen in nineteenth- and twentieth-century marxisms as limited to the context of a particular nation-state, but increasingly being analysed at a global level. It is crucial to note that hegemony is a process, not an accomplishment, that the actions of a dominant group are always open to contestation. Yet, in most societies based on the nation-state, most of the time, a relatively steady equilibrium can be observed, a state best defined as one of non-crisis, punctuated by crises that lead to the achievement of a new relative equilibrium.

For example, liberal capitalism in the overdeveloped countries of the mid-twentieth century operated under a hegemonic model of relations with the working class which was known as the Keynesian welfare state. Unions were allowed to exist and to fight for improvements in the lot of workers, in return for which corporations received a guarantee that strikes would occur only under ritualistic and tightly controlled circumstances. The state acted as an intermediary between these two hostile camps, taking money from the corporations in the form of taxes, and providing public services to both corporations and the workers. This relatively stable system stayed in place until the 1970s, when it began to be displaced by the neoliberal model, through which capitalism sought increased profits by freeing itself from the fetters of state regulation and working-class resistance. Privatization, deregulation, 'right to work' legislation (union-busting) and fanatical worship of 'the free market' became *de rigueur*. Outside the walls protecting G8 privilege, governments of countries of the global South were pushed into 'structural adjustment programmes' that had the same general thrust as in the North, but with greater intensity and much more disastrous results. Along with new national and international institutions came a new common sense: those who are oppressed deserve their oppression; everyone (except the rich) must work more for less; the bigger a corporation is the better; the less the state intervenes in the economy (except to bail out failed corporations and provide them with free infrastructure and the right to pollute at will) the better. And so on.

The shift from the Keynesian model to the neoliberal model involved a realignment of some major historical forces. Neoliberal entrepreneurs, intellectuals and journalists have been working to reverse the flow of social change, and they have been largely successful in doing so, all over the world. They have 'won the hearts of minds' of the middle classes of the global North and the elites of the global South, and they have shown their willingness to dominate—and in some cases to liquidate—antagonistic groups using armed force (union activists in Colombia, Taliban in Afghanistan, Baathists in Iraq). Neoliberalism has been seeking hegemony, and it is achieving it on a scale that makes the Chinese, Roman and Aztec Empires look parochial indeed. This is the fact we must face: capitalist globalization not only exists, it is a result of conscious planning on the part of global financial and governmental elites that meet precisely for this purpose on an increasingly regular basis. The only point worth discussing, at this juncture, is how we can best fight it.

The obvious answer is to try to establish a counter-hegemony, to shift the historical balance back, as much as possible, in favour of the oppressed. This might mean a defence of the welfare state in the global North, or a continuation of the battle to enjoy its benefits for the first time in the global South. Or it might mean attempting to establish a *different kind* of global hegemony, one that works from 'below' rather than from 'above'. To argue in this way, however, is to remain within the logic of neoliberalism; it is to accept what I call the *hegemony of hegemony*. By this I mean to refer to the assumption that effective social change can only be achieved simultaneously and *en masse*, across an entire national or supranational space. Marxist revolutionaries have followed the logic of hegemony in seeking state power, hoping to reverse the relationship between the dominated and the dominators. Liberal and postmarxist reformism display the same logic, although in a different mode—rather than seeking to take state power, they seek to influence its operation through processes of pluralistic co-operation and conflict (Laclau and Mouffe 1985; Kymlicka 1995). What is most interesting about contemporary radical activism is that some groups are breaking out of this trap by operating *non*-hegemonically rather than *counter*-hegemonically. They seek radical change, but not through taking or influencing state power, and in so doing they challenge the logic of hegemony at its very core.

In this sense many of the most interesting of the newest social movements are not what sociologists would call social movements at all. Thus there is a certain irony in my use of this term, an irony that is intended to highlight the shift away from hegemonically-oriented 'movements', and towards *non-branded strategies and tactics* such as Independent Media Centre (IMC), Affinity Group, Reclaim the Streets (RTS), Social Centre, and Black/Pink/Yellow Bloc. At the same time, however, there are tendencies that better match the sociologist's definition of social movements that also display what I call an *affinity for affinity*, that is, for non-universalizing, non-hierarchical, non-coercive relationships based and mutual aid and shared ethical commitments. Examples would include certain indigenous communities in North America (Mohawk Nation), Latin America (Zapatistas) and Australia/New Zealand (Aboriginal Provisional Government), as well as some strands of transnational feminism and queer theory. My basic argument in this book is that all of these groups and movements, strategies and tactics, are helpful in understanding—and furthering—the ongoing displacement of the hegemony of hegemony by an affinity for affinity.

WHO IS SPEAKING?

In tracking the difficult emergence of this alternative logic I will present critical readings of key thinkers from the anarchist, marxist and liberal traditions. Marx's critique of capitalist exploitation is, I believe, a deep and rich well from which we must continue to draw. Anarchist theory is equally valuable for its insistence that state domination is as great a problem as capitalist exploitation. Both of these traditions, however, suffer from certain difficulties inherent to their historical emergence in the period of high European modernity. Their commitment to social-scientific rationality and their millenarian faith in the achievement of a society entirely free of domination and exploitation have been deeply problematized by poststructuralist theory, which has produced much more complex analyses of capitalism, the state form, and the societies of control, which of course did not exist in Marx's or Bakunin's time. More importantly, poststructuralist theorists such as Foucault, Deleuze and Guattari, and to some extent Derrida have worked intensively on the question of how we might continue to struggle against oppression without reproducing the modern fantasy of a final event of totalizing change (the revolution), *or* falling back into the abyss of liberal pluralism.

While Foucault worked as a prison activist and wrote highly influential studies of sexuality, medical normalization and social control, and Derrida has addressed racism, Eurocentrism and neoliberalism in his texts, poststructuralist theorists have often been criticized for being apolitical, disengaged and blind to their own privilege (Fraser 1981; Habermas 1987; Hartsock 1990; Best and Kellner 1991). In the following chapters I will engage with these debates in greater detail; for the moment, though, I want to point out that poststructuralist influences have been felt by other traditions that are more easily recognizable as 'political'. For example, just as the liberated/Enlightened subject of classical socialism has been undermined by Foucault, Deleuze et al., so has the 'woman' whom second-wave feminisms sought to emancipate been brought into question (Nicholson 1989; Hekman 1990; Elam 1994). Postcolonial critics have carried out a similar project with respect to the Eurocentric biases of supposedly universal discourses (Minh-ha 1991; Bhabha 1994), and queer theory has challenged the heteronormative assumptions of all of the above traditions and their critics (Sedgwick 1990; Butler 1993b; Spurlin 2001). These developments have helped to push the insights of poststructuralist theory beyond its sometimes narrow perspectives, to arrive at new formulations that have important implications for the reform of existing structures. In the process, however, the radical impulse of post-1968 French theory—the impulse to create alternatives to the state and corporate forms rather than just work within them—seems to have been lost. I see myself as contributing to a small but growing body of work in postanarchism and autonomist marxism that seeks to recover this impulse, by articulating how a non-reformist, non-revolutionary politics can in fact lead to progressive social change that responds to the needs and aspirations of disparate identities without attempting to subsume them under a common project.

As valuable as the dissemination of poststructuralist critique has been, its insights are often presented in language that is accessible only to academics, and then only to those academics who have steeped themselves in some rather difficult texts and traditions. In recognition of this problem I have set out to write a book that will be of interest to activist-minded academics while remaining accessible to theoretically-minded activists. I see this effort as an attempt at creating what Gilles Deleuze and Michel Foucault have called 'relays'. In a discussion that was published under the title 'Intellectuals and Power' (Foucault and Deleuze 1976), Foucault and Deleuze reject the hegemonizing conception of intellectuals as figures who raise the consciousness of non-intellectuals, who then take practical action based on the abstract analysis with which they have been provided. For them, theory is not about abstractions, it is itself 'a struggle against power' (75–6); similarly, practice is not about simply 'applying' theoretical concepts to particular social-historical contexts. Rather, practice is 'a set of relays from one theoretical point to another, and theory is a relay from one practice to another' (74). That is, they see theory and practice as embedded in mutually interpenetrating networks that defy any attempt at separation into discrete components or moments.

At the same time, however, it is important to remember that neither 'theory' nor 'activism' exists in the kind of detached way that Foucault and Deleuze are prone to discuss them. Theory is always bound up with theoretical traditions, and activism exists only in relation to established and emerging communities. Thus, what is also at stake, as Gayatri Spivak has pointed out in her famous response to 'Intellectuals and Power', is positioning oneself with respect to, and creating relays between, disparate traditions and identities. In the case at hand, Spivak accuses Foucault and Deleuze of speaking in ways that unconsciously reveal their privileged location as professors in the French academy and their lack of solidarity with those whose existence and ability to speak/act have been obscured by European colonialism (Spivak 1988). Spivak's critique can be—and has been—generalized: those who enjoy a structural privilege must strive to identify and work against this privilege if they hope to establish relations of solidarity with those who do not share it. For this reason it is important that I identify myself as a White male university professor living and working in the relative ease and comfort of a G8 country. As an anarchist raised in a working-class family I have encountered a certain amount of disrespect and rejection by my intellectual 'superiors', but I know that my own socialization and the racist and patriarchal norms that permeate the academic world make it relatively easy for me to stay in the game. The same goes for my sexuality. Although I find that my desire exceeds the boundaries of mainstream heterosexual practices, I have a long-term female partner and two children—so again, I mostly pass. All of this to say: my struggles with oppression arise mostly from the need to challenge my own racism, heterosexism and classism, and to find more effective ways to be in solidarity with those who experience the debilitating effects of these apparatuses of division every day and night, throughout their lives. This is what I set out to do in my teaching, research and activism, and it is what guides me in writing this book.

As I have already mentioned, the systematic inequalities to which social scientists pay the most attention are not the only barriers one has to deal with when trying to work across disparate traditions of theory and practice. From the academic side, trying to avoid reproducing old prejudices means working across disciplines, and this opens one up to charges of dilution, eclecticism or dilettantism. Most academics like each other to stay on their own territories, and react with varying degrees of protectiveness when someone upsets the status quo by wandering around too much. Not only does this territorialism limit what can be said, it also gives rise to a deep prejudice as to who can say it. No one raises an eyebrow when a famous academic intellectual is quoted to support a point, but placing an equally well-established activist in that same position can get you into trouble. There are exceptions, of course, some of which are holdovers from the days when academic institutions did not have a virtual monopoly on intellectual legitimacy—Karl Marx didn't work at a university, but no political theorist would question the relevance of his work to 'the discipline'. A few contemporary writers, such as Gloria Anzaldúa, have been able to establish a presence within the halls of academia while maintaining strong connections with other communities. They have avoided becoming what Anzaldúa once called 'dependent scholars' (2000/1982: 18), but only at the cost of contending with a constant doubt regarding their qualifications and therefore the value of their work.[2] From the activist side—particularly among anarchists—there is a complementary long-standing distrust of anyone who works in an academic context. University-based researchers are often seen as parasites seeking street cred or professional advancement, or simply trying to satisfy a voyeuristic urge to participate in 'the real world'. There is also a strong distaste for academics who set out to 'teach' activists about 'theory' or 'history', that is, who assume that their own way of thinking through social and political problems is the *only* way.

For me, then, creating relays means not only challenging my own socialization and seeking ways to be more firmly in solidarity with those who struggle directly with racism, heterosexism, colonialism, capitalism and state domination. It means joining in certain academic debates while at the same time continuing to be involved in broader movements of resistance and construction of alternatives. It means providing readings of academic texts as well as attending to the voices of activists, in such a way as to minimize any signs of a hierarchical relationship between these practices. Above all, it means trying to join in existing discussions, or start new ones, while leaving myself open to the possibility that my intervention will simply bring on critical responses from all sides at once. To do as much as I can to avoid this outcome, I have tried to engage with a broad spectrum of writers, activists and movements from all over the world, within the bounds of my experience and expertise. I have tried to absolutely limit the use of jargon, and when I feel compelled to use terms that might not be well-known outside of specialized academic circles, I have provided background discussion in the main text and/or endnotes to further relevant readings. Notes are also sometimes used to address issues that are important to academic debates but may be of less interest to non-academic readers. In the prose style I have tried to walk a fine line between what those who identify primarily as academics might see as inadequately qualified statements, and what those who identify primarily as activists might see as useless belabouring of a point. In the words of Canadian activist Mathieu Dykstra: 'We need to have innovative ideas, but keep it in a language that people can understand' (Dykstra 2003).

THE ARGUMENT

With all of these concerns in mind, I will now turn to a brief outline of the argument of the book. Chapter 1 documents how the logic of hegemony is being challenged by a wide range of activist practices, which cut across all identity categories and are appearing in all parts of the world. From the refusal of work to the construction of concrete alternatives to the existing order, these dispersed and constantly morphing tactics nonetheless share some common characteristics. They are not oriented to allowing a particular group or movement to remake a nation-state or a world on its own image, and are therefore of little use to those who seek power over others, or those who would ask others for gifts, thereby enslaving themselves. Rather, they are appropriate to those who are striving to recover, establish or enhance their ability to determine the conditions of *their own* existence, while allowing and encouraging others to do the same. I argue that these affinity-based practices cannot be understood from within the horizon of (neo)liberal and (post)marxist theoretical traditions, which are dominated by the hegemony of hegemony. This domination is quite unconscious in most cases, though. The hegemony of hegemony must therefore be approached *genealogically*, as a discourse with a history that deeply conditions our present understandings and possibilities (Foucault 1985).

The following two chapters set out to provide this genealogical understanding. Chapter 2 begins by showing how contemporary marxist and liberal theorists have obscured the particularity of the newest social movements by attempting to either ignore them or incorporate them into their own paradigms. Leftish liberals such as David Held (Held and McGrew 2002), for example, have cast the more mainstream elements of anti-corporate struggle as part of an emerging 'cosmopolitan democracy', while simply dismissing radical currents as 'extremism'. The more orthodox marxists, for their part, close their eyes to the utter rejection of classical marxist politics and dream of a resurgence of class struggle (Marcuse 2000; Sweezy and Magdoff 2000b). To explain this blindness, the chapter proceeds with an account of the rise of hegemonic thinking, from its emergence in classical liberalism to its further development in Russian and Italian marxisms. I argue that, despite their many historical and theoretical differences, classical marxism and liberalism share a belief that there can be no 'freedom' without the state form (Leviathan or dictatorship of the proletariat), and therefore also share a commitment to political (state-based) rather than social (community-based) modes of social change. The paradoxical belief that state domination is necessary to achieve 'freedom' is perhaps *the* defining characteristic of the hegemony of hegemony, in both its marxist and liberal variants.

In Chapter 3 I acknowledge that the classical logic of hegemony has been marked by a series of internal challenges to the modernist assumptions upon which it was based, primarily through English-language appropriations of the work of Antonio Gramsci by postmarxist cultural studies and 'new social movements' theories of the 1970s and 1980s. Elements such as the preordained assignation of particular historical tasks to particular classes, and the primacy of working class struggles over all others, were supposedly obliterated forever in this theoretical *auto-da-fé*. Like any other adventure in deconstruction, though, this process can only go so far—certain core assumptions must be left in place if the concept of hegemony is to be recognizable as such. I therefore set out to describe carefully the limits of liberal and postmarxist pluralism. The argument begins by claiming that liberal multiculturalism, as it is practised in states like Canada and Australia, and to some extent the European Union (EU), should be seen as a paradigmatic case of what I call the *politics of demand*. This mode of social action assumes the existence of a dominant nation attached to a monopolistic state, which must be persuaded to give the gifts of *recognition* and *integration* to subordinate identities and communities. I then note that this model has been generalized, for example in academic contexts in the United States, to include not only ethnic and racial identities, but an ever-expanding set of struggles, each of which emerges out of one of the specific modes of domination and exploitation characteristic of the globalizing system—gender, sexuality, ability, age, and so on.

This is the infamous 'identity politics' that has had so much trouble establishing its pedigree against the resistance of those who see its concerns as 'merely cultural' (Fraser 1997; Butler 1998). My argument, which is supported by readings of academic texts and interviews with activists working on issues of postcolonial and queer identity, anti-racist feminism and indigenous struggles for self-determination, is not that identity politics is inadequate because it is based on symbolic constructs, but that it faces certain impediments that are inherent to the politics of demand. The most debilitating of these is the way in which this mode of action is caught up in what Lacanian theory calls an *ethics of desire*, an endless repetition of a self-defeating act that only perpetuates the conditions that give rise to its own motive force. Fortunately, the same identities that have hit the limits of the politics of demand have begun to move beyond them, towards a *politics of the act* driven by an *ethics of the real*. This alternative ethico-political couple relies upon, and results from, getting over the hope that the state and corporate forms, as structures of domination, exploitation and division, are somehow capable of producing effects of emancipation. By avoiding making demands in the first place, it offers a way out of the cycle through which requests for 'freedom' or 'rights' are used to justify an intensification of the societies of discipline and control.

This is, of course, not an entirely novel argument—anarchists have long advocated what they have called social rather than political revolution. The problem has been that the possibilities of social change without the state form have been marginalized by the dominance of (post)marxist and (neo)liberal models of social change. Another genealogy is therefore required, one that reveals an ever-present undercurrent of affinity-based theory and practice, beginning with Godwin's first tentative glimpses of how a modern non-statist society might be organized, through to the marxist critique of so-called Utopian socialism and the debates over the role of the state in post-revolutionary societies. Chapter 4 takes up this theme, by showing how classical anarchists tried to transcend the dichotomy between revolution and reform, at the same time as they struggled with racism, classism, sexism, homophobia, and their own faith in Science, Reason and even the Capitalist Market. Through these trials and tribulations anarchism developed what a theory of *structural renewal*, which begins with Godwin's notion of non-statist federalism and finds its most coherent expression in the work of Gustav Landauer (Buber 1958/1949). Through his contact with Nietzsche's work, Landauer anticipated poststructuralist theory in analysing capitalism and the state form not as 'things' (structures), but as *sets of relations between subjects* (discourses). Based on this analysis, he was able to understand how small-scale experiments in the construction of alternative modes of social, political and economic organization offered a way to avoid both waiting forever for the Revolution to come *and* perpetuating existing structures through reformist demands. Thus, although Marx and Engels were quite correct in their critique of Fourier, Owen and Saint-Simon, the tendency to identify all of anarchist theory as Utopian socialism is quite misguided. Classical anarchist theory not only moved beyond these Utopian elements, it also found solutions to problems that most marxisms still refuse to fully confront.

This is not to say, however, that classical anarchism has no problems of its own. Rather, as a number of postanarchist writers have argued, it retains the marks of its birth out of the womb of the European Enlightenment (May 1994; Newman 2001; Call 2002). Nor is marxism entirely without an awareness of the viability and value of the logic of affinity, as council communism, the surrealists, the Situationist International, and most recently the Italian autonomists have shown. Chapter 5 picks up the logic of affinity as it presents itself to us today, that is, in the context of late twentieth-century and early twenty-first-century encounters between modernist theories of radical social change and their poststructuralist critics. Against the grain of an interpretation common to both activists and academics who tend to equate US-style postmodernism with French poststructuralism, I argue that poststructuralist theory does not necessarily lead us into a zone of apolitical nihilism or pure textual play. Rather, the work of thinkers such as Michel Foucault, Gilles Deleuze and Félix Guattari should be seen as driven by a series of ethico-political commitments that defy the dichotomy between moral certainty and moral relativism. Primary among these is a commitment to minimizing domination in one's own individual and group practice, while at the same time warding off attempts at domination by others. As Rosi Braidotti (2002) has pointed out, this is what it means to work via a micropolitics, a politics of minority rather than majority, of affinity rather than hegemony; a politics that *remains political* despite its rejection of the fundamental assumptions of (neo)liberal and (post)marxist theories of social change. Dispersing and realizing this politics, however, is a non-trivial problem. Very interesting and important steps have been taken in this direction by postanarchism and autonomist marxism, each of which has its strengths and weaknesses. The non-leninist strains of autonomist marxism provide a compelling analysis of the postmodern societies of discipline and control, and pick up on the anarchist critique of hierarchical modes of organization such as the revolutionary state and party. Yet they often display a tendency towards a hegemonic totalization of the field of struggle, which appears most prominently in their conception of 'the multitude' as a singular entity organized around class struggle. Postanarchism has done better at escaping the hegemony of hegemony, but at the cost of an excessive reliance upon a 'nomadic' conception of subjectivity that appears to reject not only coercive morality, but affinity-based ethico-political commitments as well. As the complement of both the (post)marxist/(neo)liberal citizen and the postanarchist nomad, I suggest that the figure of the *smith*, as theorized by Deleuze and Guattari and as exemplified in the practices of the newest social movements, offers the greatest potential for community-based, radical social change in the twenty-first century.

If this potential is to be realized we will need not only new ways of thinking about ourselves, but also about the communities in which we live. This is the task of Chapter 6, which addresses another common misreading of poststructuralist theory, namely that it proceeds without any reference to shared understandings of social worlds. Just as the poststructuralist rejection of coercive morality has been read as a rejection of ethics and politics, here the rejection of hegemonic conceptions of community is mistaken for a rejection of community as such. This distinction is highlighted by a discussion of what Giorgio Agamben calls *the coming community*, which begins to break with the Hegelian legacy of state-based conceptions of group identity. Like the autonomist marxists with whom he is associated, however, Agamben's work suffers from a tendency to envelop singularity in the single—that is, despite his obvious commitment to multiplicity, he theorizes the coming community in a monolithic way. I suggest that we need to think instead of the coming *communities*, in the plural, but not in the form of liberal pluralism, and that we need to guide our relations with other communities according to the interlocking ethico-political commitments of *groundless solidarity* and *infinite responsibility*. In the simplest terms, groundless solidarity means seeing one's own privilege and oppression in the context of other privileges and oppressions, as so interlinked that no particular form of inequality—be it class, race, gender, sexuality or ability—can be postulated as *the* central axis of struggle. This insight has been developed most fully by postmodern/anti-racist feminist theorists, but is finding its way to other discourses and disciplines, and is gaining much currency in activist circles. The second principle, infinite responsibility, means always being open to the invitation and challenge of another Other, always being ready to hear a voice that points out how one is not adequately in solidarity, despite one's best efforts. Here, too, there are complementary currents in academic and activist practices, which have seen some successes yet must still face many obstacles. The main point I want to make in this chapter is that what we think can only be done via the state and corporate forms, through the politics of recognition and integration, can in fact be done, and done more effectively, without passing through these mediating institutions.

In the final chapter I condense the argument of the book into a concise statement of my basic thesis, which is that marxist revolutionism and liberal/postmarxist reformism have hit their historical limit, that is, the limit of the logic of hegemony and its associated politics of representation, recognition, and integration. This is an argument that will undoubtedly be controversial to those who see hegemonic practices as the only way in which they can hope to achieve the kind of social change they desire. However, as the newest social movements so powerfully show, an orientation to direct action and the construction of alternatives to state and corporate forms opens up new possibilities for radical social change that cannot be imagined from within existing paradigms. By reading the anarchist tradition critically, that is, in the light of poststructuralist, feminist, postcolonial, queer, and indigenous critiques, the value of a logic of affinity guided by groundless solidarity and infinite responsibility becomes apparent. It's time to forget the 'new' social movements of the 1960s–1980s. There's something even newer afoot, and it offers the best chance we have to defend ourselves against, and ultimately render redundant, the neoliberal societies of control.

I
Doing it Yourself: Direct-action Currents in Contemporary Radical Activism

'I shit on all the revolutionary vanguards of this planet'
Subcomandante Insurgente Marcos, January 2003

As David Graeber has pointed out in a recent article in *New Left Review*, many of today's activists are rejecting 'a politics which appeals to governments to modify their behaviour, in favour of physical intervention against state power in a form that itself prefigures an alternative' (2002: 62). This politics is radical in the sense I outlined in the Introduction, that is, it is less concerned with affecting the content of current forms of domination and exploitation than it is with creating alternatives to *the forms themselves*. In this chapter I look at a wide range of activist practices, with the intention of understanding both how they intervene against state and corporate power and how they prefigure, or in some cases create, alternatives to the existing order. For purposes of discussion I will consider an array of non-hegemonic tactics under a handful of analytic headings, arranged in increasing order of their efficacy in addressing forms rather than contents. These include: *dropping out* of existing institutions; *subversion* of existing institutions, through parody; *impeding existing* institutions, via property destruction, 'direct action case work', blockades, and so on; *prefiguring alternatives* to existing institutions, often via modes of activity that otherwise fall within the purview of a hegemonic politics, for example protests; and finally, *construction of alternatives* to existing forms that render redundant, and thereby take power from, the neoliberal project.

Some of these approaches have, of course, been used in the past under other circumstances, and none of them can be thought of as 'purely' non-hegemonic. What is at issue here is a matter of nuances, not of totalities. One particularly interesting aspect of contemporary practice is that many of the most effective tactics are *non-branded*, that is, they tend to spread in a viral way, with no one taking ownership or attempting to exercise control over how they are implemented. Unlike, say, the dictates of the Communist International in the heyday of the Soviet Union, they easily morph into new forms appropriate to different times and places, and thus are beginning to display the kind of diversity and differentiation that is required for 'survival' in the hostile environment of neoliberal societies of control.

Another important nuance is the linkage of these tactics to anarchism, which ranges from implicitly adopting traditionally anarchist methods without an awareness of their origin, to explicit display of the circle-A—which, of course, hardly makes things clear, since this emblem has been appropriated by everyone from suburban consumer rebels to software companies. I will use the term 'anarchistic' to describe (what I see as) implicitly anarchist elements in a group or tactic, reserving the term 'anarchist' for situations where there is an explicit self-identification.[1]

Finally, I should note that in researching this chapter I, along with others involved in the Affinity Project, sometimes felt as though we were working on a report for a counter-intelligence bureau. So, to minimize the chance that our work will compromise the security of any group or network, we have ensured that all of the information presented here or on our website is publicly available, or comes from interviews where the desire for anonymity has been respected.

ZERO-PARTICIPATION: CRUSTY PUNKS AND LIFESTYLE ANARCHISTS

The first current I will address is the zero-work or drop-out tactic, which is associated with surrealism, the Situationist International (SI), and, most recently, anarcho-primitivism. Bob Black's (1985) essay 'The Abolition of Work' is a strong influence here. Black criticizes feminists, liberals, marxists and most brands of anarchism for 'quibbling over the details' of work without questioning the meaning and effects of work itself. 'Almost any evil you'd care to name', he suggests, 'comes from working or from living in a world designed for work. In order to stop suffering, we have to stop working.' Although his argument is highly polemical and more than a little reductive, Black makes interesting links between the ideology of work and centralization, surveillance and authoritarianism. Classical marxists and anarchists, of course, were aware of the deadening and dangerous effects of labour under the capitalist system, but Black refuses the assumption that non-capitalist labour will be any more enjoyable. Rather, following the surrealists and situationists, he generalizes the classical socialist critique to include work in the home and school as well as the factory and office. In the background of this primitivist challenge, of course, there lurks the assumption that life need not be this way, that it has not always been this way. Thus John Zerzan argues that work as we know it is a product of civilization, an imposition upon a non-Hobbesian prehistory 'characterized more by intelligence, egalitarianism and sharing, leisure time, a great deal of sexual equality, robusticity and health, with no evidence at all of organized violence' (Zerzan 2002: 49).[2]

Anarcho-primitivist influences figure prominently within the drop-out culture that is burgeoning all over the world, but is particularly strong in inner-city areas of the United States. In an interview for the Affinity Project, Dave Battistuzzi, an organizer for the Northeastern Federation of Anarcho-Communists (NEFAC), notes that there are 'two anarchisms' at work in cities like Baltimore, where the flight to the suburbs has left behind deeply impoverished, primarily Black neighbourhoods with low housing costs and many abandoned buildings. Since the 1980s, these areas have been attractive to White-punk-DIY squatters whom Battistuzzi identifies with what Murray Bookchin (1995) has called 'lifestyle anarchism'—that is, with a 'middle-class, escapist, feel-good' sub-culture (Battistuzzi 2003). They often have no strong links to the surrounding community, and are therefore seen more as first-stage gentrifiers than as a potential source of progressive alliances. In recent years, however, a new kind of drop-out politics has emerged, which is driven more by necessity than choice, and is trying to reach out across the boundaries created by race, class and anarchist subcultures. Rather than simply capitalizing upon impoverishment and strife to acquire cheap or free housing, these squatters contribute to their neighbourhoods by providing much-needed community services. Eleanor, who calls herself 'a White anarchist kid', has lived in a number of squats in inner-city Black areas of Philadelphia. She has been involved in creating community gardens, bike workshops, free art classes and other initiatives, and has found the communities in which she has lived to be 'positive and supportive' (Eleanor 2003). Her experience offers a way out of the stark distinction Bookchin makes between social anarchism and lifestyle anarchism, by showing how living an alternative lifestyle can be combined with other tactics that are more obviously 'political' in nature.[3]

THE INCREDIBLE LIGHTNESS OF CULTURAL SUBVERSION

The past few years have seen a resurgence of another situationist technique: _détournement_, which the SI defined as the integration of 'preexisting aesthetic elements' into a 'superior construction of a milieu' (Knabb 1981: 45–6). In everyday terms, *détournement* involves taking images and text from mainstream media and subverting them for other ends. The Situationist International's journal made copious use of this technique, sometimes succeeding in its critical intent, at others failing to adequately distance itself from spectacular representations.[4] Echoing this call for radical cultural-political critique, but with a broader critical awareness, a contemporary website devoted to billboard subversion declares that '[w]e can turn the tables on capitalism if we recognize that we can all be artists—if we don't compete, but play, play hard and play seriously' (urban75 2003). Most diverted advertisements do not take on constituted powers directly, for example by proclaiming that 'Capitalism Sucks!', but work to subvert its attempts to (re)define everyday spaces, values, and subjectivities. The skilful *détournement* of a billboard, then, can be seen as a form of direct action based on the construction of situations, and can be extended beyond the critique of spectacular commodity relations to include racism, heterosexism, technophilia and the military industrial complex. Some of the more interesting work in this regard can be viewed at the website of the ironically named California Department of Corrections, which has taken up the task of remedying some of the many deficiencies to be found in state and corporate advertising.[5]

Finally, it is important to point out that, unlike their marxist precursors, today's 'situationists' do not see themselves as an artistic-revolutionary *avant-garde* with a select membership. Rather, billboard liberation is dispersed as a non-branded tactic open to all. The Billboard Liberation Front (BLF) website, for example, contains an online manual entitled *The Art & Science of Billboard Improvement* (BLF 2004a), which describes in detail the tools, methods and precautions necessary for carrying out successful actions. The *BLF Manifesto*, taking a characteristically ironical tone, makes clear the desire for wide dispersal of the billboard liberation tactic:

> Our ultimate goal is nothing short of a personal and singular Billboard for each citizen. Until that glorious day for global communications when every man, woman and child can scream at or sing to the world in 100Pt. type from their very own rooftop; until that day we will continue to do all in our power to encourage the masses to use any means possible to commandeer the existing media and to alter it to their own design. (BLF 2004b)

The Surveillance Camera Players (SCP) are also doing innovative work in cultural politics that is explicitly situationist-inspired. Like the SI, the SCP operate at the interface between everyday life and its representations within the societies of control. But they have carved out a new niche by performing live plays in front of surveillance cameras, such as their own version of the final scene of George Orwell's *1984*, Alfred Jarry's *Ubu Roi*, and original works such as the one in which they reassure watching guards that they're 'just going to work', 'just shopping', and so on.[6]

Their work is intended to be humorous and accessible, but the SCP are also engaged in a theoretically driven critique of the societies of control. 'It is the demand for and imposition of transparency', they argue, 'that unites the apparently isolated spectacle of video surveillance with the general capitalist spectacle' (SCP/NYC 2004a). Transparency here refers to the panoptic visibility of our daily lives via their representation within systems of cybernetic regulation. The most insidious aspect of this transparency, of course, is its own transparency; most of us do not notice, or do not care, that we are photographed hundreds of times in our daily round, tracked by our cellphones or located by the use of our bank cards. In calling attention to the transparency of transparency, the SCP force one of the mechanisms of the societies of control into view, rendering it susceptible—at least potentially—to critical discourse and contestation.

Like BLF, the SCP tactic is spreading virally and rapidly. A flyer used by the New York group contains the following advice:

> If you, too, are worried about the destruction of your constitutional rights in the name of 'fighting crime', we encourage you to form your own anti-surveillance camera group. You can even use the name 'Surveillance Camera Players'. (SCP/NYC 2004b)

This call has been taken up enthusiastically, with over 20 groups participating in an International Day Against Video Surveillance in September 2001, and countless autonomous actions taking place from the USA to Europe and Latin America.

While these instances of everyday *détournement* are important examples of the cultural-political battles being waged today, the struggle for control over the material-symbolic environment becomes particularly intense when an international financial meeting rolls into town. During these events, the arrival of the leaders and their entourages is preceded by months of 'site preparation', which has the goal of turning a particular human-natural community—a 'somewhere'—into a 'nowhere', a wasteland of dead power and its artifacts. Daily life is suspended as a 'safe' zone is fenced off, archaic or newly wrought laws against standing on the streets or wearing masks are trotted out, and snipers occupy the high ground.

In this kind of environment, one might expect dissent to all but disappear. However, in a concrete refutation of the thesis of a pessimistic postmodernism, the heightening intensity of repression at these meetings has been accompanied by an even greater resurgence of an aesthetic of resistance via transformative play. Colourful, mobile performances and sculptures have become increasingly common alternatives to the more traditional modes of discursive communication.

Groups like Art and Revolution and Bread and Puppets have been instrumental in perpetuating and refining this tactic, which has a lineage back to agit-prop and the work of Brecht and Jarry. Although Bread and Puppets has been guided primarily by the vision of Peter Schumann, larger events such as the Domestic Resurrection Circus (held annually from 1975 to 1998) were organized on anarchist principles:

> The organizational structure within the Bread and Puppet Theater developed in response to the requirements of the *Circus*—organically (as it were) in an anarchistic fashion, which is to say, in response to situations as they developed, with individual members of various committees taking on responsibilities as they saw fit. All met regularly with the Schumanns and other puppeteers. (Bell 1999: 108)

Like BLF and SCP, puppeteering has been disseminated as a non-branded, open tactic practised by loose networks of autonomous groups.[7]

Despite the ubiquity of this kind of spontaneous, creative, intervention, the cultural-political aspects of contemporary activism are consistently downplayed in the corporate mass media, in favour of representations of violent conflict. In the mid-1990s, Reclaim the Streets (RTS) began to receive a great deal of attention for its massive impromptu parties, held on the streets and squares of London and, most memorably, on the M41 motorway. Despite the best of festive intentions, John Jordan notes that these actions did not always succeed in their goals. 'The action which we felt failed the most was the Social Justice/Never Mind the Bollocks event—not only did it fail in that we did not manage to carry out our main plan, but also because a street party in Trafalgar Square, followed by newspaper front pages with "Anarchist Riot", "Attempted Murder", etc. is not politically effective' (Jordan 1997).

While the minions of Rupert Murdoch and Conrad Black have continued to ignore the political motivations of street parties, choosing to focus instead on their 'violent', 'irrational' and 'self-indulgent' aspects, the rapid spread of RTS implies that it possesses a certain sort of viability. What began as a tactic of a local coalition against the motorization of Britain has since been employed at the Summit of the European Council in Barcelona, the NATO meetings in Munich, and the World Economic Forum in New York City, all during the first few months of 2002. RTS is an excellent example of the dispersal of a non-branded tactic, on the basis of what the London group call 'disorganization':

> In relation to past and expected future press reports concerning trials of RTS 'leaders', Reclaim the Streets London would like to emphasize that it is a non-hierarchical, leaderless, openly organized, public group. No individual 'plans' or 'masterminds' its actions and events. RTS activities are the result of voluntary, unpaid, co-operative efforts from numerous self-directed people attempting to work equally together. (RTS London 2000a)

If this approach sounds anarchistic, that's because it is self-consciously driven, among other influences, by anarchist principles. 'The theft of time and space by capitalism, and resistance to it, along with a fusing of green (ecological), red (socialist) and black (anarchist) politics has always been central to London RTS' (RTS London 2000b).

GETTING HEAVY: IMPEDING THE FLOWS OF STATE AND CORPORATE POWER

At the same time as the RTS tactic subverts the existing order in a serious-playful way, it also provides an example of another important approach, that of impeding existing forms. Obviously, when a major motorway or downtown street is barricaded for a party lasting several hours, it becomes more difficult for the dominant system to operate as normal. This is the logic that drives an array of direct action networks, such as Earth First!, the Earth Liberation Front (ELF), and its progenitor, the Animal Liberation Front (ALF). Like RTS and the other networks discussed so far, these groups tend to favour an anarchistic mode of dispersal over centralized organization.[8]

The general principles behind Earth First! are non-hierarchical organization and the use of direct action to confront, stop and eventually reverse the forces that are responsible for the destruction of the Earth and its inhabitants. EF! is not a cohesive group or campaign, but a convenient banner for people who share similar philosophies to work under. (Earth First! 2004)

ELF is based on similar principles:

As the E.L.F. structure is non-hierarchical, individuals involved control their own activities. There is no centralized organization or leadership tying the anonymous cells together. Likewise, there is no official 'membership'. Individuals who choose to do actions under the banner of the E.L.F. are driven only by their personal conscience or decisions taken by their cell while adhering to the stated guidelines. (ELF 2004)

The ELF guidelines call for activists to 'inflict economic damage on those profiting from the destruction and exploitation of the natural environment' and to 'reveal and educate the public on the atrocities committed against the earth and all species that populate it', while taking 'all necessary precautions against harming any animal, human and non-human'. Actions in 2003 have focused on halting urban sprawl by burning down luxury housing developments under construction, and spray-painting anti-war slogans on petrol-hungry SUVs (Sport Utility Vehicles).

Examples such as these, of course, tend to give credence to the popular assumption that ecological direct action involves nothing more than mindless destruction. To avoid perpetuating this view, it is worth pointing out that among its victories Earth First! counts actions such as the successful protection of South Downs turf at Offham Hill Valley in Sussex. This intervention reversed the stereotype of eco-activism by *repairing* the damage that had been done by a farmer ploughing up land that had been in a relatively wild state. Further, I would suggest that most actions oriented to impeding flows have a constructive moment, precisely to the extent that they prevent or limit the havoc wreaked by industrial capitalism. Human private property will have little value once we have all died of cancer or radiation sickness.

Similar actions have been taking place all over the world for a very long time, so much so that any attempt to narrate a 'representative sample' cannot help but be woefully inadequate. None the less, I will mention a few more examples to give a sense of the wide dispersal, as well as the strengths and weaknesses, of this tactic. The tradition of non-violent direct action in India is particularly strong, with a well-established basis in Ghandian principles of civil disobedience. The Karnataka State Farmers Association recently drew upon this tradition to fight illegal secret trials of genetically engineered crops in India. In 1998 they launched Operation Cremation Monsanto, in which fields containing genetically modified cotton plants were uprooted and burned. 'We send a very clear message to all those who have invested in Monsanto in India and abroad,' the group proclaimed. 'Take your money out now, before we reduce it to ashes' (Kingsworth 1999). The tactic spread to other grass-roots activist groups, and for a time it appeared as though it had worked, as the Indian government halted testing. But by April 2002 the tide had turned—the Indian Genetic Engineering Approval Committee decided to allow Monsanto-Mahyco to sell the disputed seeds, at four to five times the cost of conventional hybrids (Monsanto Inc. 2002).

The famous Chipko movement worked on similar principles of non-violent direct action and autonomous, decentralized organization to protect forests and watersheds throughout India (Shiva 1988: 67–77). It was formed and led by village women who were the first 'tree-huggers'—throughout the 1970s and 1980s they placed their bodies between the forests and the saws and axes that would destroy them. In 1980 their efforts led to a 15-year ban on the felling of live trees in the Himalayan forests of Uttar Pradesh, and similar reforms were achieved in other states. The 15-year ban has since expired, however, and the story is similar to that of Operation Cremation Monsanto. At a recent meeting of former Chipko activists, it became clear that the gains that had been made were only temporary. The participants complained that 'authorities who should be helping to protect the delicate ecology of the hills are instead working hand-in-glove with the timber barons' (Dogra 2002). Now it is major hydroelectric projects and the harvesting of traditional herbs for the capitalist market that threaten the forests and the subsistence of the people who depend upon them. Some veterans of the first Chipko movement are calling for it to be revitalized, which shows both the enduring power of direct action and the necessity of endless struggle against the depredations of the neoliberal project.

Endless struggle certainly seems to characterize the direct action efforts of the indigenous peoples of Turtle Island (North America), who have been dealing with the effects of capitalist globalization for hundreds of years. Every imaginable tactic has been deployed to limit and reverse the flow of European colonialism, but certain events and the places associated with them evoke particularly powerful memories: Alcatraz, Oka, Wounded Knee, Gustafsen Lake. All of these were times when indigenous people made a stand on land that had long been part of their traditional territories, either to reclaim it or prevent its further degradation. All were times when indigenous people suffered and died in their quest, which as in other regions has seen both successes and failures. The so-called 'Oka Crisis' came to public attention in Canada on March 11, 1990, when men, women and children of the Kahnesatake Mohawk nation set up a barricade to block the expansion of a golf course onto their ancestral lands. These lands, which included a cemetery, had been in dispute for hundreds of years, but the Canadian state was showing no signs of living up to its reputation for multicultural benevolence. The Mohawk at nearby Kahnawake set up a solidarity roadblock on a bridge, which inspired the Canadian and Quebec governments to call in 2,500 troops. After an aborted attempt to break the barricade at Oka, a seven-day siege ensued, which was vividly documented in the national and international mainstream media. The golf course was not expanded, but the Oka land claim remains unresolved and tensions in both communities continue to run high.

CONAIE, the Confederation of Indigenous Nationalities of Ecuador, has had similar mixed results with regard to mainstream Ecuadorian public opinion and state policy. Over the past eight years two presidents have been brought down by indigenous opposition, which has included widespread use of direct action blockading tactics. The current president, Lucio Gutiérrez, came to power via an alliance with Pachakutik, a party representing indigenous peoples in Ecuador. Despite this alliance, the Gutiérrez administration has adopted familiar neoliberal economic and social policies, taking the side of multinational oil interests and threatening to use military force to assure the smooth flow of oil and capital through indigenous lands and lives (Pachamama Alliance 2003).

In the global North—where the oil ends up—there is also a well-established tradition of direct action outside of indigenous groups, from IWW (Industrial Workers of the World, or Wobblies) tactics of workplace sabotage in the early part of the twentieth century, to Quaker anti-nuclear protests in 1958 and of course the anti-globalization protests of the late 1990s (Kubrin 2001). This activity has continued into the 2000s with the US dockworkers' blockade of major west coast ports in solidarity with the Seattle protest, and various actions to impede the operation of the US war machine both in the 'homeland' and abroad. But the most recent and spectacular example of direct action to impede the flows of state and corporate power would have to be the Black Bloc tactic, which has almost become obligatory at major convergences. By participating in a Bloc, activists offer up their semi-protected bodies to state-sponsored violence, in the hope not only of saving other protesters from physical harm, but also to provoke shock, horror and perhaps even dissent among liberal citizens who hold to values like freedom of speech and the right to legitimate protest . Also, with their balaclavas, garbage can lids and baseball bats, Black Bloc members offer a parody of the riot police, and thereby threaten the legitimacy of the monopoly of state and corporate forms on the use of violent force to attain their ends.[9]

Perhaps most subversive of all, though, is the challenge that the Black Bloc tactic offers to the monopoly on invisibility and silence, with its active ignorance of the command not only to behave well, but to be available to be *seen* behaving well. In refusing to follow the rule of transparency which guides the societies of control, Bloc subjects represent glaring exceptions within the domesticated and privileged strata of the global North. Not only has the system of cybernetic regulation failed to modulate their behaviour properly, but they also seem to be immune to self-discipline, fear of physical punishment, and verbal and physical attacks by other activists an academics.

For all of these reasons, it is surprising that we have heard and seen so much of the Black Bloc in the mass media. It must be remembered, though, that the best way to ensure the exclusion of a radical social force is to ensure its *inclusion* within the spectacle, where its meaning can be appropriately modulated (witness the fate of communism modulated as social democracy). In the case of the Black Bloc, this is done by way of an appeal to the time-honoured tropes (discussed in the introduction) that play on a fear of anarchist outsiders bent on the destruction of civilization as we know it. Encountering these representations, the liberal citizen is first moved to fear; but the fear need not last, for he is immediately able to see that the state and the corporations are taking care of him by responding with violence to the imagined threat created by young men and women 'armed' with garbage can lids—a beautiful *détournement* if ever there has been one! One can also imagine, though, that the violence of subjects in revolt is put on general display so as to avoid the much greater dangers inherent in images of peaceful, playful, subjects advancing meaningful critiques. On this analysis, the spectacular recuperation of the Black Bloc tactic would stand as a corroboration of one of the central claims of situationist theory—perhaps spontaneous, joyful construction of situations *is subversive after all*, and that's why the Black Bloc tactic is burgeoning into all the colours of the rainbow, including the queer anti-capitalist Pink Blocs now making regular appearances and sometimes marching alongside their Black counterparts—but so far without the same attention from either the mass or the alternative media.

No discussion of impediment of flows of power in the globalizing societies of control would be complete without mentioning attacks on the informational infrastructure. Although it is obviously limited in its applicability to regions where internet access is affordable and the required expertise available, hacktivism is becoming an increasingly important form of direct action (Jordan 2002). This tactic emerged in the 1990s, as the tools created for raising awareness of internet censorship and surveillance were turned to other uses. Deploying concepts developed by Deleuze and Guattari, the Situationist International and Hakim Bey, two books by the Critical Arts Ensemble (CAE) provided the theoretical impetus for this development: *The Electronic Disturbance* (1994) and *Electronic Civil Disobedience* (1996). In the latter text, CAE argue that as the state and capital have gone postmodern—fluid, electronic—civil disobedience has remained attached to its modernist roots, attacking buildings and supposed centres of power:

> These outdated methods of resistance must be refined, and new methods of disruption invented that attack power (non)centers on the electronic level. The strategy and tactics of CD can still be useful beyond local actions, but only if they are used to block the flow of information rather than the flow of personnel. (CAE 1996: 4)

Although CAE's almost total dismissal of the efficacy of blocking material flows is decidedly Eurocentric—and has been proven by recent events to be incorrect even in the global North—groups such as the Electronic Disturbance Theatre (EDT) have shown that electronic civil disobedience can be a potent tactic, either on its own or allied with material actions. Denial of service attacks against WTO and FTAA websites during the Seattle and Quebec protests, for example, duplicated on the virtual level what was happening in the streets. The Floodnet programme, developed by EDT, mechanized the denial of service tactic, so that small numbers of activists were able to slow or shut down a target website. However, there are limits to the efficacy of hacktivism, which are already becoming apparent. The results of virtual attacks are not necessarily visible to the general public or clearly attributable to political rather than merely technical causes; and being one of a few people sitting at home alone in front of a screen just doesn't seem to provide the same sense of solidarity and empowerment as facing down the police in the streets of a major city on a hot summer day.

DIRECT-ACTION CASEWORK: A HYBRID FORM

While it is relatively easy to argue that dropping out, dancing and puppetry are merely personal and selfish pursuits, or that burning crops, breaking windows or crashing computer systems is just plain old vandalism, it is more difficult to critique the next mode of direct action I want to discuss: interventions on the field of constituted power on behalf of marginalized individuals and groups. This is the strategy that guides the activities of the Ontario Coalition Against Poverty (OCAP), which operates in Toronto, Canada. According to OCAP's founder John Clarke, 'as a militant, anti-capitalist organization, we reject the notion that we have any set of common interests with those who hold economic and political power. We also reject the rituals of token protest that confine movements to the level of futile moral arguments ...' (Clarke 2001). One of their more successful events was the 'Dave's Discount Supermarket Action', which occurred in 1995 at a major chain food store in downtown Toronto. In response to a 20 per cent cut to social assistance rates and Social Services Minister David Tsbouchi's suggestion that welfare recipients who could no longer afford to feed themselves should haggle for discounts, 50 OCAP members filled their grocery trolleys at the store, then attempted to pay for their purchases at checkout counters using 'Dave's 21.6% Discount' coupons. The havoc created in the store drew considerable media attention to the welfare cuts and highlighted the callous ridiculousness of the minister's remarks.

On request, OCAP also intervene on behalf of particular individuals, in a mode they call 'direct-action casework'. In May 2002 they were approached for help by a woman who had been denied social assistance because she refused to give up a volunteer position with an NGO, even though this position was to lead to a full-time paid position in a short period of time. OCAP sent a letter of complaint to the welfare office, threatening public direct action if the grievance was not resolved. The agency responded by providing assistance to the woman, without requiring that she leave the volunteer position. OCAP counted this a victory based on the efficacy and high profile of their tactics : 'It is clear how OCAP's well-established readiness to use direct action methods to confront such injustices has by now created a situation where the mere threat of a response often brings results' (OCAP 2002).

Over the years OCAP has been successful in stopping deportations, cancelling evictions and forcing employers to pay back-wages, leading to their tactics being adopted by other groups, such as the No One Is Illegal [NOII] campaign based in Montreal. This campaign has focused on the plight of 'illegal' refugees in Canada, some of whom are facing incarceration and deportation, especially with the tightening of immigration laws after the Day of the Great Excuse for Oppression, that is, September 11, 2001. NOII activists recently succeeded, via direct action against immigration offices and officers, in blocking the deportation of Algerian families who face persecution and death in their home country (Ahooja and Schmidt 2003). NOII activists have pointed out, though, that the members of the one of these families, who were involved in a high-profile sanctuary action, 'spoke excellent French, had never been on welfare, and had worked throughout their ... stay' in Canada (10). That is, the action was able to sway public opinion partly because it appealed to the model of a 'racist, classist, and imagined normative Canadian citizen' (10).

Mujeres Creando (Women Creating), an anarcha-feminist collective operating out of La Paz, Bolivia, are known primarily for their graffiti, street theatre, and video and television work (Paredes 2002). The focus of their work is on 'deconstructing machismo, anti-gay prejudice, and neoliberalism' (Ainger 2002: 107), through actions that are guided by a rejection of the politics of parties and states:

> We decided on autonomy from political parties, NGOs, the state, hegemonic groups who wish to represent us. We don't want bosses, figureheads, or exalted leaders. Nobody represents anybody else—each woman represents herself (Ainger 2002: 107).

In 2001, members of the group began working with a group of small debtors who wanted relief from a corrupt World Bank-financed 'micro-credit' scheme that had left many of them bankrupt and subject to repossession of the few belongings they might have had.

They helped to organize a number of events, ranging from non-violent local direct action to workshops on neoliberalism. After camping in La Paz for over three months and getting no action, the debtors decided to take over the Defensoria del Pueblo (People's Defence) office, the office of the Catholic archbishop, and the agency that was responsible for supervising the banks involved in the micro-credit scheme. At the bank, they tied up 94 executives and demanded negotiations. Although Mujeres Creando had not been involved in organizing the occupation, women from the group were brought in to act as mediators between the occupiers and the state and corporations, a role they took on partly to 'prevent a massacre from taking place' (Julieta Ojeda, in Styles 2002). They were successful in brokering a peaceful end to the occupation, in which the Bolivian government promised to investigate how the financial institutions were implementing the micro-credit scheme and put off property seizures for 100 days. The debtors were not successful, however, in achieving their main goal, which was to have the debts cancelled.

This outcome, considered alongside the efforts of NOII and OCAP, highlights the fact that direct-action casework contains elements of protest, and thus can be said to be driven to some extent by reformist goals. It proceeds by asking figures in positions of structural power to take notice and change their actions. However, tactics like this will certainly be necessary to help ward off the re-emergence of capitalism and the state form within spaces that have been liberated from the neoliberal project, and their value in achieving practical results, here and now, should not be downplayed. If one more family is allowed to penetrate the fortress of the global North, if one more forest is saved from destruction, then there are greater possibilities for more radical forms of social change, or simply a greater ability to survive until such forms take deeper root.

Direct-action casework also shares a limitation common to all actions that seek to impede the flows of state and corporate power: while they may be successful in the short term in particular cases, over the long term and in the majority of cases, the impeded flow tends to find another outlet. One forest isn't cut, but another is; one family isn't deported, but dozens are denied entry to avoid further disruptions to the immigration system. This problem is inherent to direct action to impede flows and will not go away. However, if this kind of action proliferates sufficiently, the flows overall will start to decay beyond the ability of systems of control to manage them. This is especially true as the neoliberal world order expands in size and complexity. Because it is hierarchical and centred (yes, even in the era of 'decentralized corporations'!), it becomes more fragile as it grows. Extending this line of analysis further, though we encounter another problem: the sudden collapse of the neoliberal order would indeed create the conditions for a modernist revolution, which many of us would find quite heartening. But, as has happened so many times before, very few people would be ready to accept a life of non-domination and non-exploitation—most would seek new masters, and a few would try to accommodate them. Avoiding the quest for masters requires some experience in alternatives to slavery; it requires prefiguration of other ways of being within and alongside existing practices.

PREFIGURING/CREATING ALTERNATIVES

The meaning and value of prefiguration are consciously discussed and widely accepted by contemporary radical activists, such as John Jordan of London RTS:

> RTS does not see Direct Action as a last resort, but a preferred way of doing things ... a way for individuals to take control of their own lives and environments .... If global capitalism does not manage to destroy the ecosphere and human civilization ... and a new culture of social and ecological justice is developed, RTS would hope that direct action would not stop but continue to be a central part of a direct democratic system. (Jordan 1997)

A similar line is taken by Stephanie Guilloud, an anti-globalization organizer involved in the Battle of Seattle:

> In the streets, we relied upon trust and consensus to make our quick decisions about how to respond to tear gas and where to move next. Our process embodied the nonhierarchical vision we were working to realize. (Guilloud 2001: 226)

The form upon which Guilloud and the others relied was the affinity group, which has become ubiquitous in the late 1990s and 2000s. It emerged in Spanish anarchist circles in the late nineteenth and early twentieth centuries, where it was adopted in conscious opposition to hierarchical marxist styles of political organizing. 'A movement that sought to achieve a world united by solidarity and mutual aid', Murray Bookchin notes apropos of the *grupo de afinidad* Solidarios, 'had to be guided by these precepts; if it sought a decentralized, stateless, non-authoritarian society, it had to be structured in accordance with these goals' (Bookchin 1998: 180). Formed out of a shared desire to accomplish a specific task, affinity groups are consensus-driven and oriented to achieving maximum effectiveness with a minimum of bureaucracy, infighting and exposure to infiltration. They tend to be small, typically consisting of 5–20 individuals. Affinity groups have formed the basis for successful actions carried out across a broad spectrum of engagement, from decentralized service groups such as Food Not Bombs (FNB), to AIDS activists ACT UP and the clandestine cells of the Earth Liberation Front. They have also become a favourite organizing tool at major anti-globalization convergences, where the model has been extended to larger groups of groups via the mechanisms of clusters and spokescouncils.[10]

Some activists see the affinity group as a form that is most appropriate for actions that are illegal or otherwise can't be public. For larger-scale organizing, more open and inclusive groups might better suited to bringing in new members. But it seems clear that affinity groups are 'good for developing personal dynamics, for dealing with issues like sexism and racism' within social movements, and that for major convergences, the spokescouncil is 'something we all understand', a method that allows people from different regions and ideological perspectives to come together to implement a common vision (Battistuzzi 2003). The affinity group, then, is not an organizational panacea. But as I will argue later in this book, it is a model that can be, and has been, extended to larger groups and non-statist federations. Certainly it must be remembered that its value lies not only in achieving political efficacy and organizational efficiency, but in building alternative cultures and societies— alternative subjectivities and ways of being—within the currently hegemonic order.

The Temporary Autonomous Zone (TAZ) is another widely adopted tactic for bringing people together in novel ways, though by its nature it is much less structured than the affinity group/cluster/spokescouncil model. The TAZ concept was developed by Hakim Bey, as an alternative to the fading dream of totalizing revolution:

Are we who live in the present doomed never to experience autonomy, never to stand for one moment on a bit of land ruled only by freedom? ... Must we wait until the entire world is freed of political control before even one of us can claim to know freedom? (1991b: 98)

The TAZ, as its name implies, is always intended to live a short life, but an intense one. It is 'like an uprising which does not engage directly with the State, a guerilla operation which liberates an area (of land, of time, of imagination) and then dissolves itself to re-form elsewhere/elsewhen, *before* the State can crush it' (1991b: 101). The notion of the TAZ has been enthusiastically taken up by the Rave scene (Thomassen 2002), but also shows up in contexts that are more overtly and consistently radical and political. Indeed, it could be said that many of the tactics I've discussed here involve the creation of TAZs, in the form of momentarily reclaimed streets, summit convergences or occupations to block environmental destruction. As I will argue in Chapter 5, the TAZ concept also holds promise for more permanent modes of association that not only prefigure alternatives, but actually create them.

Once again, it needs to be stressed that I am not claiming that the construction of alternatives to the state and corporate forms is an entirely novel phenomenon. For far longer than states and corporations have existed, individuals and groups have been self-organizing to autonomously meet their own needs (Kropotkin 1989/1902). The oldest and most prevalent way in which this is done is via informal community networks, which underlie even the most 'advanced' of liberal-capitalist societies, yet are constantly displaced as modernity—and now postmodernity—further their colonization of what Jürgen Habermas calls the lifeworld (1984). That we are never very far from being able to meet our own needs is made apparent by the way in which people of the global North respond to events such as major blackouts or snowstorms. When deprived of the good life delivered by the grids, we immediately leave our living rooms (there's nothing on television, so what the hell) and head outside to see what's going on. Very quickly, friends and neighbours find ways to keep themselves warm and fed by working together and sharing what they have, and only in situations where class and racial antagonisms are determining does 'anarchy', or disorder, break out—most of the time, what we see is the best of anarchism. Of course, when the power comes back on everyone retreats once again into their 'private' lives. But in situations of continued crisis, these kinds of relationships can be incorporated into long-term daily life.

One example of this effect can be found in Chile after the US-orchestrated military coup that ousted the government of Salvador Allende. The new regime, which was supposed to rescue the country from the evils of socialism, was massively repressive and brought in economic policies that devastated the country, hitting the poor the hardest. In order to feed their families, women set up *ollas communes*, or communal kitchens, to buy, prepare and serve food in their neighbourhoods. The idea soon spread throughout the country, prompting state repression and making it necessary for many of the kitchens to go underground or keep moving from house to house (Hensman 1996: 50–1). Argentina has also became a hotbed of spontaneous informal organization when the economy collapsed at the end of 2001. Utterly disgusted with years of state terrorism and the disappearance of their savings while those of the rich were safely transferred offshore, thousands of Argentinians began to rally around the slogan *'Que se vayan todos'*, 'Get rid of them all'. As in Chile, this was not a movement of a few hard-core activists, but one that took in vast segments of the lower and middle classes across the country; at one point, it is estimated that over 200 *asambleas populares* (popular or neighbourhood assemblies) existed in Buenos Aires alone (Project Censored 2004). The assemblies have taken upon themselves the role usually handed to the states and corporations, by setting up communal kitchens, participating in the administration of hospitals, organizing protests and direct-action interventions, and working in solidarity with the *Piqueteros* (Garrigues 2002). It is, of course, not only the outcomes that are important here, but the process itself both prefigures and creates alternative political structures based on direct democracy, community accountability, and individual/group empowerment. As one member of a neighbourhood assembly puts its:

> We are creating a community in the desert, in the desert of the big city where looking someone in the eye is difficult. Security used to be in the bank, and insecurity was in the streets. Now insecurity is in the bank. The robber who used to be outside the bank is now in it. And security is in the streets, with our neighbours. (Pablo 2004)

The Argentine state has, of course, responded to this threat to its legitimacy and efficacy with surveillance and violent repression, despite the peaceful and constructive nature of their activities. Some assemblies have been colonized by traditional left-wing parties, which seek to tap their power and redirect it into their hierarchical, authoritarian structures. Despite these efforts, the *asambleas* continue to exist in Argentina, and have begun to spread to Venezuela and Brazil. In 2003 an 'Autonomista Caravan', organized by activists from

Brazil and Argentina, travelled from Montreal to Miami, spreading the word about these grass-roots initiatives in North America (Huff-Hannon 2003). The struggle between community (not civil society!) and state and corporate forms is indeed *the* struggle of the (post)modern condition.

Community networks for self-support can also be formalized by way of the creation of co-operatives. Granted, the co-op movement has fallen on hard times in most parts of the world. Many building societies in the UK have been dismantled through demutualization schemes that disperse collective capital to individual shareholders, and neoliberal states in the global North have done everything they can to cripple, download and outright eradicate the co-operative sector. As always, their analysis is sound: every co-op protects a space of autonomous activity that is potentially colonizable by profit-seeking capital, and every co-operative subject is a little less docile than he or she would otherwise be. But in regions where the tradition is deeply rooted, and especially where fears of cultural genocide spur the quest for social, political and economic autonomy, co-ops remain strong and vibrant. The theoretical basis for the co-operative movement can be found in the work of the Utopian socialists Charles Fourier and Robert Owen (see Chapter 4). They provided the models for self-sustaining local communities, which were applied by themselves, in Owen's case, or by others. The Rochdale Society, formed in 1844 in England, is often seen as one of the first fully functioning co-operative experiments, and provided a model for the Canadian and US movements of the early twentieth century (Pybus 2003: 56). The co-operative sector never really took off in the US, but remains strong in many parts of Canada, especially in Quebec or where there are significant Quebecois communities. Co-ops and other economic collectives have also been used extensively by women in the global South, who are able to occupy economic niches that are compatible with their status as primary care-givers for their families and which make use of skills that are traditionally associated with 'women's work' (Apena 1995–96; Carr et al. 1997; H. Klein 2001).

In addition to the time-honoured, though always threatened traditions of formal and informal community organization, the past few years have seen a remarkable proliferation of new groups and networks, once again best seen, I think, as non-branded, non-hegemonic alternatives to the neoliberal project. The burgeoning network of Independent Media Centres (IMCs) is an excellent example of this kind of 'productive' direct action.[11] IMC aims to combat corporate concentration in media ownership through the creation of alternative sources of information, and in so doing to participate directly in the negation and reconstruction of mass-mediated realities. Not only is each centre independent from the corporate world, it is also independent from the other centres—there is no hub which disseminates a particular editorial line, and on some parts of some sites, there is no editorial line at all. Each centre tends to be driven by the interests and resources of the local communities it serves, thus building a high degree of differentiation into the system at its most basic level. Again, what makes this tactic important in the context of social movements is its political logic, as the following account from a participant-researcher involved in the Vancouver, Canada IMC makes clear:

> Independent Media Centre is, I think, one of the most important recent examples where grassroots movements, particularly those in the North, work to create spaces that are autonomous from capital and the state, where processes unfold according to logics dramatically opposed to the instrumentalist logics of accumulation and centralized decision making, even while these movements use technologies created for these purposes. It is also an instance of a subtle shift in political activism and struggle, a move from strategies of demand and representation to strategies of direct action and participation. (Uzelman 2002: 80)

In a proposal for a Zapatista-style *encuentro*, a coalition of IMC activists highlight issues that are relevant to any network that hopes to avoid the slide into centralization and rational-bureaucratic domination that has plagued so many hegemonically-oriented groups: 'how to build open, inclusive, decentralized structures of accountability, decision-making, and action locally, regionally, nationally, and globally?' How to 'bridge gaps in gender, colour, culture, age, access, language and "otherness" for capacity building and empowerment?' (IMC Encuentro Proposal Working Group 2000). The IMC tactic is just one example of how decentralized, reconstructive communities can reclaim media spaces. Others include infoshop, pirate radio and television, and, of course, 'zines and paper pamphlets. This is one area where the master's tools have been of some use in bringing down his house.

The non-branded tactic model has been adopted by many other networks that are reclaiming spaces and lives from the states and corporations. The first FNB chapter came into existence in 1980 in Cambridge, Massachusetts, formed (as the name implies) by anti-nuclear activists working under the Clamshell alliance, which included anarchists, Quakers and marxist socialists (Food Not Bombs 2004; Werbe 1999). Its mode of operation is simple: take food that will be wasted, use it to make vegetarian meals and serve them for free to people who are hungry. Anyone who has spent time in an inner-city park or attended a protest or alternative music festival in the past ten years has likely tasted the benefits of this extremely productive form of direct action. Like the IMC and RTS networks, FNB is a non-branded, decentralized network of autonomous chapters which function internally on a consensus basis. Unlike the IMC tactic, though, FNB is easily transportable outside of the fortress of first-world privilege, and has spread to every continent, with affiliated groups in Turkey, South Africa, Australia, Argentina and India, to name just a few.

The social centre is another very interesting non-branded tactic that emerged out of Italy in the 1970s. In the early years of this decade the Italian revolutionary left had been forming *comitati di quartiere,* or neighbourhood councils, as a complement to the workers' councils which formed the basis of their organizing strategy. Since their community had no pre-school, medical clinic or library, militants from one of these committees in Milan 'occupied and reactivated' an abandoned building, and invited the newly elected city council to 'demonstrate in practice its intention to meet the social needs of the population of a popular neighbourhood like ours, allowing for the social use of the occupied factory' (Leoncavallo Occupation Committee 1975). The Leoncavallo centre, as it came to be known, seems to have been an ideological melting pot: anarchists set up a printing press, theatre groups presented shows in a tent adjacent to the building and radical feminists created a women-only space or *Casa Delle Donne* (Cimino 1989). There was also a carpentry school and a *Scuola Popolare* (People's School) which allowed workers to acquire a middle (secondary) school education. Soon other centres began to appear in Milan, and eventually all over Italy, some of which—like Leoncavallo—have survived to the present day, despite state repression and fascist assassinations of their members.[12]

The movement in Italy today, according to some, has lost much of its original character, including its militancy and its intimate contact with the working class. This has been a combined effect of repression, changes in styles of militancy, and a strong presence of non-revolutionary punk-anarchist elements in the Italian social centre movement. 'Those [centres] that work', argues one former social centre activist, 'are integrated into the metropolitan society of the spectacle ... in a certain way they function as a business, internally they are like a cooperative that organizes shows and offers them to the public for a certain price in order to finance themselves but also in order to stay in the market.' 'For this reason', he continues, 'we think that the CS [social centre] is no longer a form of struggle to pursue in itself for itself' (CSA 2003). Not everyone in Italy shares this analysis however, and the model has been enthusiastically adopted elsewhere. ABC No Rio is a long-standing New York social centre which began in 1980 with an occupation of a vacant city-managed property, an action dedicated to Elizabeth Mangum, a Black woman who had been killed by police as she resisted eviction in Flatbush the previous year. The occupation was carried out to obtain space for an art exhibit called 'The Real Estate Show', which sought to demonstrate that 'artists are willing and able to place themselves and their work squarely in a context which shows solidarity with oppressed people, a recognition that mercantile and institutional structures oppress and distort artists' lives and works, and a recognition that artists, living and working in depressed communities, are compradors in the revaluation of property and the "whitening" of neighborhoods' (Committee for the Real Estate Show 1980).

Social centres are opening up all over North America, sometimes via squatting, but also on a rental or donation basis. This is the case with Project 1877 in Pittsburgh, which I visited shortly after it opened in 2003. It describes itself as a 'community space connecting activists', but as a local alternative newspaper reported, it has become 'a community space connecting the community' (Eldridge 2003). It houses the local IMC, does shows and films catered by the local Food Not Bombs chapter, and runs a free space for local street artists to work. Aspire is a relatively new collective that has carried out a series of squats/social centres/autonomous zones in Leeds (UK) since 1999. They run a vegan café, children's activities, a DIY art space as well as various shows and film nights. The Radical Dairy is a similarly mobile squat/centre operating out of London, and is allied with the London Social Centre Network, which includes a half-dozen centres and seeks to 'link up the growing number of autonomous spaces to share resources, ideas and information' (London Action Resource Centre 2003). Similar spaces are popping up throughout Europe, Australia and New Zealand, which shows that while most individual sites may come and go quickly, the social centre tactic is here to stay.

The occupation of space to prefigure and create autonomous alternatives is not limited to privileged subjects of the global North. Argentina's Movimento de Trabajadores Desocupados (Unemployed Workers Movement) have been extremely successful in using the tactic of highway blockades to express frustration with existing institutions. In August 2001, for example, a nation-wide mobilization of federated local groups managed to close more than 300 highways, thereby severely limiting the ability of the capitalist economy to maintain the flow of goods upon which it depends. The *Piqueteros*, as they are commonly known, are un- and underemployed Argentinians who have taken to direct action as their jobs have disappeared and their unions have been broken. Like the other groups discussed here, they use non-hierarchical forms of autonomous organizing:

> We understand by autonomy the ability which, as a people, we have to organize ourselves and 'direct' ourselves by our own selves. Our movements are independent of the State and its institutions, political parties, and the church, but autonomy goes beyond that independence: we reject the subordination of popular organizations to any superstructural petition, since we believe that the people organizing themselves from the grass roots within their own areas should be those that determine, in a democratic way, the decisions and the politics that follow. (Unemployed Workers of Lanus et al. 2003)

The *Piqueteros* do not limit themselves to blockades, however. Like the neighbourhood assemblies, they have become involved in many different projects to meet local needs, including bakeries, organic gardens, clinics, and water purification. They have also developed links with, and helped to inspire, hundreds of occupations of abandoned factories throughout Argentina, in industries ranging from textile manufacture to ceramics. Once again, this method of creating alternatives to state and corporate control has spread to North America, as 550 employees of an Alcan Aluminum plant in Quebec seized, occupied and ran the plant themselves after it closed in early 2004. In Argentina, the occupations have, of course, met with violent repression, while in Canada the attacks took the form of a labour board injunction (Engler and Mugyenyi 2004). Yet they continue to occur, as they offer the only way of meeting the needs of communities that have been abandoned by the states and corporations.

The Movimento dos Trabalhadores Rurais Sem Terra (MST, Brazilian Landless Peasants Movement) of Brazil has become well known—and thoroughly vilified by the Brazilian corporate mass media—for its role in organizing occupations of unused farm land. It is 'an autonomous movement, independent of the political parties' and the Catholic Church, but with links to the Brazilian Workers' Party (PT) and a history of engagement with liberation theology (Stedile 2002). Taking advantage of a clause in the Brazilian Constitution that allows unused agricultural land to be settled and brought into production—which was itself a victory of the movement—the MST first educates farmers about their rights, then provides guidance and support for any locally organized occupations. A crucial aspect of the MST's tactics is that in each case the farmers organize and equip themselves, so that they build the kinds of skills and relationships that are necessary if the action is to succeed in the months *before* the occupation. It is partly for this reason that 250,000 families have occupied and gained title to 15 million acres land, and have built 60 food co-operatives and a number of small agricultural industries and schools (MST 2003).

But it must be noted that state support also had a lot to do with success of the MST's tactics. Not only was the legal basis for the occupations provided by the constitution, but the state-funded Agrarian Reform Institute (INCRA) also provided money to help establish those who expropriated land, including low interest credit for cooperatives. This all ended with the coming to power of the Cardoso government in 1999, which opted instead for a neoliberal, World Bank-supported scheme for 'market agrarian reform' (Petras 2000). Rather than supporting autonomous communities and individuals, this plan will create a new class of heavily indebted farmers, desperate and presumably ready to begin cultivating Monsanto's latest monster-crop. Successive governments since 1989 have been doing everything they can to undermine the work of the MST, from raids, arrests and torture to wrapping up key figures in court battles So, although 75,000 people are currently squatting on unused land or on roadsides, one long-time movement leader acknowledges that 'for the last two years we've made very few gains' (Stedile 2002: 93–4). The MST was always a hybrid project in terms of its political logic, using direct action to bring about legal, political and economic reforms. As the state turned against it, under pressure from the global neoliberal order, it has found itself increasingly on the defensive.

Struggles over land have also been a constant theme of post-apartheid South Africa. Despite the achievement of state power by

the African National Congress (ANC) in 1994, 85 per cent of the
land remains in the hands of the White minority. Although there
are vestiges of marxist social democracy in some policies, all levels
of government are following the neoliberal agenda in the hope of
securing a better place in the global order, and this is leading to
frustration with, and cynicism about, formal political organizations.
As Stephen Greenberg puts it, 'political democratization has opened
up spaces for organization, but has also institutionalized struggles'
(2004: 27). Frustrated by the repeated failure of locally-based actions
to achieve meaningful reforms in land tenure over the 1990s, a
number of grass-roots groups came together in 2001 to form the
Landless Peoples Movement (LPM). LPM describes itself as a 'national
movement', supported by a network of NGOs organized as the
National Land Committee. The LPM, however, is striving to maintain
itself as a 'completely independent grassroots structure of landless
people. It is not an affiliate of the NLC' (LPM 2001). LPM activists
have supported 'informal settlements', or occupation of unused land,
but tend to see this as a tactic oriented to forcing the state to live up to
its promises, rather than as prefiguration of autonomous alternatives
(Greenberg 2004: 29–30). In a recent election boycott campaign,
however, there is evidence of an increasing disenchantment with
reform-oriented strategies and tactics:

> We are sick and tired of being used as pawns by political elites who only
> 'care' about us at election time, then expect us to suffer our poverty and
> dispossession in silence for the next five years. We do not believe that going
> to the polls will do anything to change our lives. Only direct, organized action
> by the united masses of poor and landless people across South Africa can
> solve the land crisis that has made us slaves in the country of our birth.
> (LPM 2004)

In the past two years the South African state—as well as some ANC
members—has responded by arresting large numbers of LPM activists
under dubious circumstances, in one case charging them with murder.
It remains to be seen whether the LPM will be further radicalized by
this response or be swept back into the NGO/state-corporate nexus
that dominates the South African political landscape.

BEYOND REFORM, THIS SIDE OF REVOLUTION

RTS, IMC, neighbourhood assembly, Social Centre, Food Not Bombs,
land and factory occupation—all of these tactics consciously defy

the logic of reform/revolution by refusing to work through the state, party, or corporate forms. Instead, they are driven by an orientation to meeting individual/group/community needs by *direct action*. Not only do they refuse to deploy traditional tactics that seek to alter/replace existing nodes of power/signification, their own organizational structures are designed so as to avoid situations where one individual or group is placed 'above' others in a hierarchical relationship. Many of these formations are aware of, participate in and support each other's activities and struggles—many squats see themselves as TAZs, are allied with IMC and FNB groups, and link up with local struggles against gentrification, racism and police brutality as well as helping to mobilize regionally, nationally and internationally for anti-globalization education and activism. Many of the same people work under different banners at different times, without facing charges of 'incorrectness' or 'going over to the enemy'. Networks like People's Global Action and Via Campesina, and convergences such as the World Social Forum are working globally to make the same kinds of links.

These organizations, and others like them, will be discussed in greater detail in Chapter 6, as prefigurations of the coming communities, or non-corporate, non-statist federations. In this chapter my goal has been limited to showing that many of the most vibrant elements of contemporary radical activism are driven by a common political logic that escapes the categories of traditional social movement theories. Unlike revolutionary struggles, which seek totalizing effects across all aspects of the existing social order by taking state power, and unlike the politics of reform, which seeks global change on selected axes by reforming state power, these movements/networks/tactics do not seek totalizing effects on any axis at all.[13] Instead, they set out to block, resist and render redundant both corporate and state power in local, national and transnational contexts. And in so doing, they challenge the notion that the only way to achieve meaningful social change is by way of totalizing effects across an entire 'national' or 'international' society. That is, they are undoing the hegemony of hegemony that guides (neo)liberal and (post)marxist theory and practice. Taking this project further in the appropriate theoretical contexts is the task of the next chapter, which examines the attempts of these dominant traditions to 'understand'—that is, to co-opt and domesticate—the newest social movements.

2
Tracking the Hegemony of Hegemony: Classical Marxism and Liberalism

> There is no longer the division between reform and
> revolution, not because the reasons for either have
> disappeared, but because the political traditions behind
> these concepts have exhausted themselves.
>
> 'CSA', former Italian social centre activist

In order to understand precisely how the logic of hegemony is
being challenged by certain elements of contemporary radical social
movements, I will now turn to a discussion of some of the ways in
which academic commentators have tried to understand these activist
currents. I don't pretend that what I will present here is anything like
a complete overview of the relevant positions. Rather, I will draw from
selected writers who exemplify certain broader tendencies within the
liberal, neoliberal, marxist and postmarxist traditions. My goal in
each case is to assess the ability of these paradigms to comprehend
what is 'newest' about contemporary radical social movements. In so
doing, I will engage not only with the current configurations of the
dominant political paradigms, but with the historical developments
that have led them to become what they are.

As theorists of hegemony have long pointed out, dominant ideas
tend to take on an appearance of naturalness and inevitability that
renders them relatively impervious to critique. This is precisely—and
ironically—what makes the hegemony of hegemony so difficult to
talk about, and even more difficult to escape. But, like every other
discourse, hegemonic thought does have a history, and this history
can be critically examined to show how it forecloses alternative
understandings of the past, present and future. To work a history in
this way is to work *against* it, to refuse to accept the basic assumptions
that allow it to function. It is to move away from history as such,
towards a genealogical account that offers new narratives with new
kinds of social, political, and economic relations in mind (Foucault
1985). In the case at hand, the goal is to show how the logic of

hegemony has become hegemonic, how it has come to structure the political sense that is common to (neo)liberalism and most forms of marxism, including postmarxism. At the same time, I want to show how the theory and practice of hegemony are unravelling, being taken apart from within their own traditions by the very forces that had to be excluded to establish these traditions in the first place.[1]

In the Introduction I proposed a preliminary definition of hegemony as a struggle for dominance, generally limited to the symbolic, geographical, economic and political context of a particular nation-state or group of states, but increasingly occurring at a global level. This definition was an attempt to capture the shades of meaning that this term evokes in postmarxism, cultural studies and other disciplines of the humanities and social sciences. Such an attempt always fails, of course, so I will now begin to unpack this definition, to give it life by placing it in its historical contexts.[2] Like so much in the western tradition, the concept of hegemony originated in Ancient Greece, where the term *hegemonia* signified the domination of one city-state by another.[3] The rhetorical content of this term is not apparent from the dictionary definition, however. To understand this we must note how it is used with reference to what is commonly presented as the commanding height of Ancient Greek civilization: democratic Athens, which provides a mythical foundation for western ideas about freedom and equality. Athenians are thought to have had a 'natural' impulse to govern themselves, but the scholarly literature is full of references to 'The Spartan Hegemony' and 'The Theban Hegemony', that is, to 'exceptional' times when (rich, genetically correct, male) Athenians were governed by others. Similarly, Philip of Macedonia (Alexander's father) is known for having established himself as the *hegemon* (leader) of most of Greece, primarily by way of superior military force. Thus to be hegemonized meant *to be unable to rule oneself* because one was under the sway of another; not another class, or even another nation—Spartans, Thebans and Macedonians were all considered Greeks—but another *political formation* in which one did not have an equal voice. Hegemony, in Ancient Greece, was very clearly seen as a *non-democratic* from of political organization.

In its current usage the concept of hegemony is deeply tied up with the system of nation-states that began to form with the rise of European constitutional monarchies, and was further entrenched by the creation of institutions of liberal democracy. Thus, hegemony must be seen as very much a *modern European* phenomenon. Its conditions were established by Enlightenment liberals, who did not use the term as such, but who provided later theorists with a rich array of concepts that were essential to the appearance of *gegemoniya* as a key term in the debates between Russian socialists of the late nineteenth and early twentieth centuries. To support my contention that the logic of hegemony deeply structures the two leading traditions of western social and political theory, I will now turn to a more detailed discussion of how these two traditions—commonly thought to be mutually incompatible—in fact share a basic set of assumptions about social organization and social change that deeply structure—and severely limit—their ability to comprehend contemporary radical social movements.

WE CAN'T HEAR YOU! LIBERALS AND MARXISTS ON THE NEWEST SOCIAL MOVEMENTS

As I noted in the Introduction, the corporate mass media have a strong tendency to ignore, vilify or denigrate any practice that poses a serious challenge to the neoliberal order. One can only expect so much (very little in fact) in the way of awareness of history and current events from these sources, but surely the bar should be set higher for leading academic intellectuals. In his book (with Anthony McGrew) on *Globalization and Anti-Globalization*, the political scientist David Held sets out to provide his readers with a discussion of the 'key political positions in favour of, and against, globalization' (2002: ix). The tone of the analysis is guarded, but the authors' position is apparent in the decision to include a great deal of discussion of whether anything like globalization is really happening at all—a move which may strike some as approaching holocaust denial. Indeed, they wonder whether those who believe that there are 'important transformations going on in the spatial organization of power' do not tend to 'exaggerate [the] scale and impact' of these changes (120–1). In clearing the way for their own position, Held and McGrew are dismissive of both 'radicals' and 'neoliberals', each of whom they regard as caught up in this kind of extremism. Instead, they focus their attention on what they see as an 'overlapping ground of cosmopolitan social democracy' (131). Picking up on the language of liberal multiculturalism, they suggest that 'this common ground in global politics contains clear possibilities of dialogue and accommodation between different segments of the "globalization/anti-globalization" political spectrum' (131).

A total of four pages of the book are devoted to what Held and McGrew call 'radicals', a strange amalgamation of anarchists, participatory democrats, marxists and communitarian liberals, all supposedly driven by 'New Left ideals' (114), and all 'wildly optimistic about the potential for localism to resolve, or engage with, the governance agenda generated by the forces of globalization' (130). Although the attempt to lump together these disparate traditions is theoretically stunning, it makes a certain sort of sense if one is working from a very large distance, as Held and McGrew clearly are. Into this category they have simply placed everyone whom they think is neither a neoliberal nor a social democrat.

Not all is lost with respect even to these esoteric currents, however. To the extent that the World Social Forum meetings have displayed 'a new emphasis on working with, and the reform of, the UN system', Held and McGrew find the participants to be in a position of productive 'overlap' with their preferred form of cosmopolitan social democracy. Unfortunately, however, there are some 'radicals' who 'do not seek common ground or a new reconciliation of views', such as 'various anarchist groupings and those notorious for attacking Starbucks at the 1999 Seattle WTO meeting' (115). Held and McGrew are quite right, of course, in noting that anarchists do not seek reconciliation with neoliberalism, but they are ill informed or disingenuous in claiming that they do not seek common ground with anyone at all. It would be helpful to have this, and other points, clarified, but unfortunately these two sentences are all that the authors find necessary to dispense with anarchist tendencies—indeed, any tendencies that are radical in any accepted sense of the term—in anti-globalization activism. Thus they not only repeat the stock tropes of the capitalist mass media, but fail in their social-scientific duty to respect empirical reality—a sin of which professors should be deeply ashamed. But they are in good company here, for, despite their pretence to an inclusive universalism, liberal-democratic theorists of globalization have an alarming habit of ignoring radical voices of dissent.[4]

Marxist scholars and activists have also been paying quite a lot of attention to the anti-globalization movement since Seattle. Unlike the liberal theorists, they advance a radical critique of globalizing capital—no holocaust denial here—which is to say that they are very much aware of the links between anti-globalization and anti-capitalism. Leo Panitch's *Renewing Socialism* (2001), for example, presents a hard-hitting analysis of neoliberalism as part of an attempt to reconstruct the (marxist) socialist project for the new millennium.

At the same time, he explicitly acknowledges the failures of actually existing socialism, and wonders what, at the end of the twentieth century, 'could be said to remain of the [marxist] socialist project' (1). He is able to see that the groups that came together to create People's Global Action (PGA) are anarchistic in their approach, in that they are committed to decentralized, non-hierarchical forms of organization (181). He also shows an awareness of the shift towards direct action, in noting that networks like PGA are not 'putting forward a series of demands that can be negotiated within the given institutional frameworks of globalization' (179). In an uncommon gesture for a committed marxist, he even admits to being impressed by the 'sight of steelworkers declaring solidarity with anarchists on the streets of Seattle' (179).

It seems clear that Panitch's vision of a new socialism is driven by an honest attempt to grapple with contemporary realities, including the rise of the newest social movements. At the same time, however, he remains committed to most of the central tenets of marxist socialism. The 'revolutionary possibilities of the working class' (10) are his central concern, and he believes that 'the salience of class will have to be brought more centrally back to the analysis and strategy of the Left' (11). Explicitly following the lead of Marx and Engels in the *Communist Manifesto*, he stresses the need for a 'new type of socialist internationalism' (11), based on 'strategically coordinating economic decision-making' (6). For him, rather tellingly, 'the key long term condition for an alternative to globalization is democratic investment control within each state' (182). Thus Panitch's 'alternative to globalization' appears to be a kinder, gentler sort of capitalism, tamed by state control—hardly a renewal of marxist socialism, and hardly a goal shared by the anarchistic elements of radical social movements today.

A special issue of the marxist journal *Monthly Review* (52:3, 2000) displays a similar ambivalence. This issue is devoted to discussing the anti-globalization movement which, like Panitch, the editors see as heralding a 'new internationalism'. They note, accurately, that Seattle was not the first 'large militant protest' against the policies of the WTO, IMF and World Bank. These had been going on for many years in the global South, where they escaped the notice of the mass media in the G8 countries. But they do give a certain pride of place to Seattle, in that it refuted the 'carefully cultivated, widely projected image of the United States as hegemonic power lacking internal social contradictions' (Sweezy and Magdoff 2000a: 1–2). That is, it gave lie to the illusion perpetuated by people like Held and McGrew, not only as it applies to the United States, but as it applies to the neoliberal project in general.

It as at this point, however, that another sort of illusion comes to the fore—the editors of *Monthly Review* see Seattle as evidence of 'the partial revival of [a] labour movement that is finally showing signs of attempting to chart a new course', that is 'rising phoenix-like from the ashes (Sweezy and Magdoff 2000a: 2). True to their tradition, and directly after insisting that the working class is leading the way, Sweezy and Magdoff administer their own kind of corrective: '[W]e are immediately faced by the reality that much—in the United States *most*—of this new wave of protest, insofar as it takes an articulated form, is directed at corporate globalization rather than global capitalism' (3). The article then turns to an analysis of 'the laws of motion of capitalism in our time', based, of course, on *The Communist Manifesto*. This is not to say that the classical marxist critique of capitalist political economy was off the mark or has been entirely surpassed—it was not, and it has not. But what the editors of *Monthly Review*, like Panitch, call a 'retreat from class' (Sweezy and Magdoff 2000b: 1) should perhaps be seen as a necessity of history's great march forward, rather than a deviance to be lamented. The same goes for Peter Marcuse's subtle attempt, in the same issue, to defend state domination by 'dispensing with the myth of the powerless state and avoiding the fallacy of the homogeneous state' (2000: 27). It is quite possible to be as critical of the state form as one is of capitalism, while holding the state to be neither powerless nor homogeneous—one simply needs to see these apparatuses, in an Althusserian way, as overdetermined components of a system that exceeds both of them.

CLASSICAL LIBERALISM AND THE BOURGEOIS REVOLUTIONARIES

Although marxists have long been known for their revolutionary politics, they were not the first to get hold of the idea that the way to achieve desirable social change was through taking state power. Credit for this innovation must go to the bourgeois activists of the English, American and French revolutions. As Hannah Arendt has noted, the 'Glorious Revolution'—that fixed the notion of sudden, violent social change in the western imaginary was, in fact, a restoration of English monarchical power after its usurpation by Oliver Cromwell. Revolution originally meant what it sounds like it *should* mean—a return of the same, a repetition or recovery (1977/1963: 43). Similarly, Arendt notes, the French and American revolutionaries were also attempting to recover a way of life that they saw as having been cast aside by absolute monarchy in one case, and colonial domination in the other. This history of the term is of more than etymological interest, since it relates very clearly to one of the central metaphors of liberal theory, which began to emerge around the same time as the idea of revolution. I am referring here to the so-called 'state of nature', the way of being that represents the ultimate return from what we might call the state of the state.

Thomas Hobbes, writing during the English Civil War and in the midst of the seizure of Paris by anti-absolutist forces, began his political deliberations with the assumption that 'men [*sic*] have no pleasure (but on the contrary a great deal of grief) in keeping company, where there is no power able to over-awe them all' (1996/1651: 88). Although he consistently came down on the side of awesome power, he was not so clear about which power was most worthy of our submission. Throughout the *Leviathan*, and prior to its writing, he appeared to be strongly on the side of absolute monarchy: 'And as the Power, so also the Honour of the Sovereign, out to be greater, than that of any, or all the Subjects. For in the Sovereignty is the fountain of Honour' (128). But at the end of the book he calls for submission to the side that he thinks has won the civil war in England, that of the parliament and the republic, reversing his earlier position. Why the about-face? Perhaps because, while he expected earlier on that the monarchy would be restored, at the time he wrote the conclusion to his book Cromwell's victory appeared secure. Hobbes seemed to believe that one should always submit to a conqueror, once one had been conquered. But conquest, for Hobbes, is not merely a military operation—in fact, one can only conquer *oneself*:

> Conquest, is not the Victory itself; but the Acquisition by Victory, of a Right, over the persons of men [*sic*]. He therefore that is slain, is Overcome, but not Conquered: He that is taken, and put into prison, or chains, is not conquered ... But he that upon promise of Obedience, hath his Life and Liberty allowed him, is then Conquered, and a Subject. (485)

Why would anyone do this? Hobbes answers: because we fear the 'natural' power of each other, we choose to submit to a 'civil' power that rules us all. This power that allows not only for pleasure, but for

Industry, Culture of the Earth, Time, Arts, Letters, and Society itself, Hobbes called the Leviathan, which we know as the modern state.

Curiously, given the generative powers claimed for it, the state so conceived does not in fact give anything to its citizens; rather it takes away something that should be rather precious—their ability to govern their own lives. Because of this paradox, it is crucial to liberal ideology that the transfer of individual autonomy to a coercive state apparatus be seen as based on consent, that it take the form of a 'contract'. In the words of John Locke, 'wherever therefore any number of Men [*sic*] are so united into one Society, as to quit every one of his Executive Power of the Law of Nature, and to resign it to the public, there and there only is *Political or Civil Society*' (1988/1690: 325; italics in original). Thus, from its earliest conception, liberal freedom requires a supposedly voluntary disavowal of individual autonomy.

The way in which Locke is using the term civil society is very different, it must be noted, from the usage of later theorists, in that it includes the legislative and executive powers of the state (325). In the early works of liberalism, civil society was not seen as opposed to, or even differentiated from, political society; rather, both terms were used synonymously to distinguish the realm of society from that of nature. 'Civil' in this context means something more like 'civilized' than it does 'private'. However, I believe the point I am trying to make does not stand or fall on the way in which the state apparatus, human society, nature, economy and polity are theorized by classical liberalism. What is crucial is that through the creation of a split between the individual (who becomes a citizen) and his or her autonomous activity (which becomes subject to state regulation), liberal citizens come to believe that *there can be no freedom without the state form.*

The highest expression of this belief can be found in the work of G.W.F. Hegel, who was also caught up in the revolutionary upheavals of his time.[5] Hegel acknowledges, in the *Phenomenology of Spirit*, that two contradictory views are possible on the question of the relation between the individual and the state. The individual may see state power as 'an oppressor and the Bad', since the state 'disowns action qua individual action and subdues it into obedience (1977/1807: 303). In Hegel's bourgeois vision, the opposing principle is 'Wealth, which the individual is likely to see as the Good'. But he points out that this judgement arises from a condition of incomplete reflection, an 'ignoble' and 'self-centred' position (305). Upon achieving a 'noble' point of view, the individual comes to 'see in public authority what is in accord with itself ... and in the service of that authority its attitude towards it is one of actual obedience and respect' (305). The pursuit of wealth is of course not entirely given up, and state power at this stage of reflection remains devoid of what Hegel called Spirit, which might best be thought of as participation in an overarching cultural unity. But it is clear that, for Hegel, the state and capitalism are essential elements for the achievement of the higher stages yet to come.[6]

Indeed, he argued that individuals are indebted for their very existence to 'the universal sustaining medium, to the *might* of the entire nation' (213; italics in original) as it is expressed in the state form. At the same time, however—and this is in keeping with Hegel's dialectical method—that 'being-for-self' is surrendered 'as completely as in death' to dominant national structures and processes, 'yet in this renunciation [it] no less preserves itself' (308). Hegel thus provides an altruistic motivation for the simultaneous abandonment of individual autonomy and its recovery as freedom within the nation-state, an act that Hobbes and Locke were able to justify only on the grounds of ignoble self-interest. As marvellous and ecstatic as this union might sound, it must be pointed out that it is in conflict with another important aspect of what Hegel says it means to be social—that is, our mutual recognition of each other as beings with will and autonomy, our participation in a *Sittlichkeit*, or ethical community.[7] Hegel adopted a series of assumptions, passed down from Hobbes and Locke, which holds that there must be a state; that this state must be used to coerce the citizens of an associated nation into a pseudo-consensual form of universal communion; and that the only alternative to this pact is to be abandoned to a nasty, brutish life of deadly competition over the means of bare subsistence. It is therefore not surprising that liberal theorists have consistently ignored anyone who suggests that a life without the state might be superior to one within it, while liberal states and corporations have had a tendency to ruthlessly crush or incorporate any group or movement that seems to be proving the validity of this suggestion through its practical activity.

HEGEMONY = DICTATORSHIP + DEMOCRACY

Marxist governments, of course, have also been extremely adept at ruthlessly crushing autonomous forces. Compared to their liberal counterparts, however, Lenin, Stalin and Mao were relatively honest and open about their activities, and justified them theoretically. How could it be that marxist socialism, as *the* ideology of freedom in community, became its horrible opposite? Was this the result of a string of bad luck, of pursuing 'socialism in one country', or was it, as Bakunin pointed out in the 1880s, a predictable result of the revolutionary strategy of Marx and the marxists? It is well known that Marx and Engels set out to stand Hegel on his feet, that is, to bring German idealism down to earth by inverting its conception of where the 'ground' of history actually could be found: in material relations rather than in ideas. But, as is always the case with deconstructive critique, the body that they subjected to this acrobatic treatment retained some of its key features. From Hegel, Marx and Engels took the idea that struggle between antagonistic forces is fundamental to historical development. But, where the liberals saw this as a battle between isolated individuals, the 'scientific socialists' framed it as one between antagonistic principles brought to earth. 'The history of all hitherto existing society is the history of class struggles', they famously declared in the opening salvo of the *Manifesto of the Communist Party* (Marx and Engels 1888/1848: 40). For Marx and Engels, history occurs upon a stage that Hobbes and Locke characterize as the state of nature, and ends with communism, as the achievement of a properly 'civil' social order. That is, the same narrative that provides a mythic origin in liberalism provides a mythic endpoint in marxist socialism. This is possible because Marx and Engels, again like Hobbes, Locke and Hegel, believed that it is 'only in community' that we may find our freedom (Marx and Engels 1978/1848: 197). They add, however, that the bourgeois conception of community is inadequate because it is not sufficiently universal: 'freedom has existed only for the individuals who developed within the relationships of the ruling class and only insofar as they were individuals of this class' (197). Thus, liberal community is 'illusory', and needs to be displaced by a socialist community that is 'real' (197).

In attempting to characterize this community, Marx and Engels also adopted other key concepts from liberalism, including its conception of civil society. 'The form of intercourse determined by the existing productive forces at all previous historical stages,' they declared, 'and in its turn determining these, is *civil society*' (1978/1848: 163; italics in original). This sphere, as defined in 'The German Ideology', 'embraces the whole commercial and industrial life of a given stage and, insofar, transcends the State and the nation, though, on the other hand again, it must assert itself in foreign relations as nationality, and inwardly must organize itself as State' (163). Marx and Engels' understanding of the relation between civil society and the state differs from Hegel's in that the latter, as we have seen, attempts to resolve the contradictions between them through an immersion in Absolute Spirit. Marx and Engels are careful to leave civil society in its place, as it were, explicitly ascribing to it lasting, determining, as well as determined, relationships. Their account is also more complex than that which was taken up by many marxists, in that it notes the determining effects that civil society (the superstructure) can have on 'the existing productive forces' (the economic base), as well as the effects that proceed from base to superstructure. However, it remains the case that Marx and Engels saw civil society as necessarily expressed in/expressive of a system of nation-states, at least until the end-time has come and history has produced its final flower, the classless society.

How are we to arrive at this union with Absolute Community? The same way that the bourgeoisie have arrived at their Holy Land, that is, by violent revolution. Marx and Engels were quite impressed with the ability of this ascending class to impose its will upon others, and were particularly appreciative, though ironically so, of its role in ridding Europe of absolutism:

> The bourgeois, historically, has played a most revolutionary part ... wherever it has got the upper hand, [it] has put an end to all feudal, patriarchal, idyllic relations ... . In one word, for exploitation, veiled by religious and political illusions, it has substituted naked, shameless, direct, brutal exploitation (1888/1848: 44–5)

The bourgeoisie, as a 'false' universal class, had achieved much. Now, using the same methods of violent usurpation of state power, the proletariat was expected to 'create a world after *its* own image' (47; italics added). It would raise itself to the position of ruling class, and thereby, as the only truly universal class, 'win the battle of democracy' (74).

In this extremely cursory treatment I do not claim to have solved, or even to have adequately engaged with, any of the numerous debates within marxist theory that are relevant to the points I have made. To do so would be desirable from an intellectual point of view, but would be a diversion from my core argument, in which Marx and Engels appear primarily as a pivot point between Hegelian liberalism and state socialism. Their work will be further engaged when it becomes necessary to show how their position was based on an explicit foreclosure of other extant options, namely, those offered by the Utopian socialism that they found so dangerous. But for now I must move on to a discussion of those who came after them—the Russian marxists, who first used the term hegemony in the context of modern social and political theory.

In his *Prison Notebooks*, Antonio Gramsci contends that credit for the 'great event' of the 'theorization and realization' of hegemony should go to 'Illich', that is, Vladimir Illich Lenin (Gramsci 1971: 357). Debates abound as to Lenin's relation to the theory of hegemony, Gramsci's relation to Lenin and leninism, and Lenin's relation to an earlier generation of Russian socialists that includes such figures as G.V. Plekhanov and Pavel Axelrod.[8] Once again, I cannot hope to provide a definitive resolution to these debates, and have no interest in trying to do so. What I hope to show, rather, is how a certain understanding of the development of the theory of hegemony has become dominant within contemporary western marxism, and those branches of academic theory and activist practice upon which it has had an enduring influence. This understanding holds that, since at least the time of Lenin, marxist theories of hegemony have contained a dual aspect of consent and coercion.[9] While the consensual aspect of hegemony remained secondary in Lenin's theory—and certainly in his practice—the dominant understanding holds that it was further developed by Gramsci. For my argument, what is most important is the claim that Gramsci's development of the consensual aspect of hegemony represents not only a *significant* advance beyond leninism, but an *adequate* advance for the understanding and furtherance of contemporary radical social movements. Refuting this claim is the central task of this chapter and the one that follows.

As Perry Anderson has shown in an excellent article in *New Left Review* (1976), the concept of hegemony (*gegemoniya*) was first deployed in a socialist revolutionary context by Plekhanov and Axelrod. At this time in Russia, the energy of political forces of all stripes was focused on the struggle against Tsarist absolutism. Plekhanov was among those who felt that the coming revolution would of necessity be bourgeois in character, as had been the case in Europe. The best the proletariat could do, he argued in the 1880s, was to echo the demands of the bourgeoisie, and wait patiently until Russian society had advanced to the point that a revolution became feasible (Anderson 1976: 15–18). By the late 1890s, however, Axelrod began to argue that the proletariat could, and must, lead the way to radical social change. All of these discussions were based, of course, on the assumption that the class that 'led' a revolution would become the ruling class in a new order to be established—it was completely taken for granted that there would continue to be a Russian nation-state; what was at stake was who would be predominant within this formation. Hence Martov declared that '[t]he struggle between the "critics" and "orthodox" Marxists is really the first chapter of a struggle for political hegemony between the proletariat and bourgeois democracy' (cited in Anderson 1976: 16). Hidden within the concept of hegemony then, from the very moment of its emergence in marxist theory and practice, was a set of assumptions that narrowed the possibilities of organizing both revolutionary struggles and post-revolutionary societies. In Chapter 3 I will discuss the context in which these assumptions were made, and give voice to the possibilities that were silently foreclosed, through a discussion of the relationship between Marx and Engels and the Utopian socialists. But for the moment I want to continue with the genealogy of hegemony, turning now to the way in which this concept was handled by the person who would eventually lead the revolution that defied marxist orthodoxy—Lenin.

In the pamphlet *Reformism in the Russian Social-Democratic Movement*, published in 1911, Lenin took up Axelrod's position and argued that 'as the only consistently revolutionary class of contemporary society, [the proletariat] must be the leader in the struggle of the whole people for a fully democratic revolution ... The proletariat is revolutionary only in so far as it is conscious of and gives effect to this idea of the hegemony [*gegemoniya*] of the proletariat' (Lenin 1963/1911: 232–3). Similar statements can be found throughout his writings, for example in *Two Tactics of Social-Democracy in the Democratic Revolution*: 'We propose to lead (if the course of the Great Russian Revolution is successful) not only the proletariat, organized by the Social-Democratic Party, but also [the] petty-bourgeoisie, which is capable of marching side-by-side with us' (1975/1905: 40). Where were Lenin and the proletariat going to lead the Russian people? To a 'revolutionary–democratic dictatorship of the proletariat and the peasantry' (52). While these claims clearly show Lenin's commitment to the line established by Plekhanov and Axelrod,[10] they beg the question of precisely how the proletariat was supposed to attain hegemony while at the same time preserving democracy. Not being one to leave loose ends dangling, Lenin provided a detailed answer to this question in what is certainly his most widely known work, *What is to be Done?* (1967/1902).

Taken in its historical context, *What is to be Done?* must be seen as a precisely directed polemic against 'spontaneity' and 'primitivism', two tendencies within Russian revolutionary circles that Lenin found particularly dangerous. In debates at the time, the 'spontaneous' activities of the revolutionary classes—strikes, reading circles, destruction of machinery—were contrasted with the 'methodical' activities of professional revolutionary intellectuals—writing letters and pamphlets, public speaking and defeating the efforts of the police to break up the movement. Those who valued spontaneity held that a successful revolution could be carried out by a proletariat that created, from within its own ranks, its own strategy and tactics. Lenin, on the other hand, argued that 'the spontaneous struggle of the proletariat will not become its genuine class struggle until this struggle is led by a strong organization of revolutionaries' (1967/1902: 132). For him, working-class spontaneity was valuable only to the extent that it could be harnessed and directed by external forces. His critique of primitivism follows directly from this assumption; if they are successfully to lead the spontaneously awakening masses, revolutionary intellectuals must become 'professionals' (107), that is, they must specialize in their chosen activity, pay close attention its theory and history and 'all of the rules of the art' of revolution (108). To proceed otherwise would be to attempt 'the kind of warfare ... conducted by a mass of peasants, armed with clubs, against modern troops' (99).

Motivating Lenin was a strong desire to resuscitate what he saw as the glory years of Russian revolutionism. 'If you are not amateurs enamoured of your primitive methods, what are you then? ... do you think that our movement cannot produce leaders like those of the [eighteen] seventies?' (104). What the leaders of the 1870s understood above all were the exigiencies of working in the autocratic context of Tsarist Russia, with its secret police, press censorship, incessant raids on revolutionary organizations, sham trials, and so on. Indeed, Lenin argued that 'secrecy is such a necessary condition for this kind of organization that all other conditions (number and selection of members, functions, etc.) must be made to conform to it' (133). His ideas regarding the need for a tight formal organization and centralization of the professional revolutionary cadres were well known, and he had already been attacked for his anti-democratic tendencies. But democracy implies publicity, Lenin argued in response, and a secret organization could not, by definition, operate in public. Nor, for the same reasons, could its leaders be elected. Thus, in an autocratic country, 'broad democracy' in revolutionary organization becomes a *'useless and harmful toy'* (136; italics in original).

To avoid a facile and one-sided critique of vanguardist modes of organization, it is important to note that Lenin made a distinction between the forms appropriate to the cadres of professional revolutionaries and those which were applicable in the organization of the workers themselves. 'Centralization of the secret functions of the *organization*', he pointed out, 'by no means implies centralization of all the functions of the *movement*' (122; italics in original). He believed that workers' groups that sought to engage a 'broad public' should be 'as loose and as non-secret as possible' (123), and should 'remain without any rigid formal structure', so as to minimize their chances of being infiltrated (115). Thus Lenin represents social-democratic revolutionary intellectuals as being forced into an unpleasant but necessary exigency—they value democracy very highly, but are pragmatically constrained to work in anti-democratic ways because of the repressive context in which they find themselves.

If this were the case, that is, if Lenin believed that non-democratic forms of organization were appropriate only to absolutist contexts, then one would expect a hundred flowers to have bloomed, as it were, *after* the revolution. But instead we find him, in 1919, arguing against those 'hypocrite friends of the bourgeoisie', those 'stupid dreamers' who believe that 'pure' democracy can be achieved at once, without passing through a phase of dictatorship of the proletariat (1955/1919a: 21). Now it is not the secret police who make it impossible for democracy to flourish, but the need to 'suppress the resistance of the exploiters', that is, of the remnants of the capitalist class within Russia. By 1923, the last year of his active political life, Lenin had begun to point out that external capitalists also represented a threat to the nascent Soviet revolution. 'It is not easy for us ... to keep going until the socialist revolution is victorious in more developed countries', he lamented (1966/1923: 498). Invoking once again the theory of proletarian hegemony, he declared that '[w]e must display extreme caution so as to preserve our workers' government and to retain our small and very small peasantry under its leadership and authority' (499). The way to do this was to make the state apparatus more efficient; as efficient, indeed, as the countries of western Europe. But, once again, the problem of spontaneity reared its ugly head. Lenin felt that the workers who were 'absorbed in the struggle' to build a new order were 'not sufficiently educated. They would like to build a better [state] apparatus for us, but they do not know how.' Education, of course, could make up for this lack, but Russian primitivism, once again, stood in the way: 'we have elements of knowledge, education and training, but they are ridiculously inadequate compared with all other countries' (488). So, Lenin proposed tightening the requirements for those workers who were recruited to the Central Control Commission—they would need to be professional bureaucrats who could 'pass a test in the fundamentals of the theory of our state apparatus, in the fundamentals of management, office routine, etc.' (491).

Taking the analysis presented in *What is to be Done?* in the context of Lenin's later interventions, it would seem that his mistrust of democratic forms in 1902 cannot be wholly attributed to the difficulties of operating under Tsarist absolutism. Once that pretext had been eliminated, he simply found other enemies who rendered democracy a useless and harmful toy. This leads one to wonder if he didn't believe, despite his rhetoric, that authority and bureaucracy were necessary for social order as such. There is in fact ample evidence for this assumption. In *What is to be Done?* Lenin closes the section in which he calls for a secret, centralized, professional revolutionary organization with an example from the experience of English trade unions. In the beginning, he notes, the members of these organizations 'considered it an indispensable sign of democracy for all the members to do all the work of managing the unions' (1967/1902: 138). Jobs were rotated, all questions came to a vote, etc. But this didn't last. 'A long period of historical experience was required for workers to realize the absurdity of such a conception of democracy and to make them understand the necessity for representative institutions, on the one hand, and for full-time officials, on the other' (138). So it wasn't Tsarism, it wasn't the internal remnants of the capitalist class and it wasn't what Stalin came to call the problem of 'socialism in one country' that drove Lenin to authoritarianism and bureaucracy; it was his belief that *no social order at all* could exist without top-down control, which meant the hegemony of the proletariat over all other social classes, and the hegemony of various professional cadres over the proletariat.

This is, of course, not a new critique of Lenin or leninism. A detailed discussion of his theory of revolutionary organization is important to this genealogy, however, as it helps us to understand how he uses the concept of hegemony to deftly unite what would otherwise be seen as two antithetical terms—dictatorship and democracy:

> The Soviet system provides the maximum of democracy for the workers
> and peasants; at the same time, it marks a break with *bourgeois* democracy
> and the rise of a new, epoch-making *type* of democracy, namely, proletarian
> democracy, or the dictatorship of the proletariat (1966/1921: 54)

In *Report on the Right of Recall* he acknowledges that the capitalist state
is an institution through which the 'entire people' is coerced by a
'handful of plutocrats'. Under conditions of proletariat dictatorship,
however, he says that it is possible to use the state to ensure 'the
carrying out of the will of the people' (1955/1919b: 13). Liberal
theorists, of course, will recognize this formulation immediately as
the one that justifies the Leviathan. Thus it is with Lenin, I would
argue, that revolutionary marxism began its long march to liberal
postmarxism. Curiously, this was in a certain way a trip back in time;
that is, postmarxism can be seen as an engagement, in the context
of the overdeveloped countries of the 1980s, with the question faced
by Russian revolutionaries of the 1880s: how are we to complete the
bourgeois revolution? But in going so far back I reach too far ahead.
The link between Lenin and Laclau must first be established by a
reading of Gramsci, who took the concept of hegemony and adapted
it for use in the 'democratic' states of western Europe.

After Lenin's death, the meaning of the term hegemony continued
to shift. In the debates of the Third International, it began to be
used to refer not only to the need for the proletariat to exercise
leadership over the bourgeoisie during a hypothetical revolution,
but to describe the position of the bourgeoisie over the proletariat in
actually existing capitalist societies. It was via his involvement in the
Comintern that Gramsci became familiar with the debates around
the theory of hegemony. Fascist Italy was hardly a model of liberal
tolerance, but Gramsci's intervention in these debates is based on
the deployment of what we have seen is a core metaphor of liberal
and marxist political theory—the division of modern societies into
the realms of the state and civil society.

As I have noted, this division has a long history, and it mutates
once again with Gramsci.[11] In the *Prison Notebooks*, he suggests that
contemporary western societies contain two major 'superstructural
levels': one that can be called 'civil society,,' that is the ensemble of
organisms commonly called 'private'; and another that he refers to
as 'political society', or 'the State' (1971: 12). Gramsci claims that
his notion of civil society is taken from Hegel's *Philosophy of Right*
(1977/1807), but as Joseph Femia (1981) has pointed out, by seeing

both the state and civil society as 'superstructural' levels, Gramsci seems to consider them to be analytically separable from the 'base', that is, from the economy. This is in contrast to Hegel—and to common liberal-capitalist usage—in which the 'private sphere' is identified very strongly with capitalist economic activity. But it is important to note that in Gramsci this separation is a product of a particular analytic point of view. Taking a different point of view, he notes elsewhere that 'between the economic structure and the state with its legislation and coercion stands civil society' (Gramsci 1971: 208). This passage seems to imply that the state, civil society and the economy are interlocking yet relatively autonomous systems, *dialectical* systems in the Hegelian sense.[12]

In the sphere of civil society, Gramsci argued, the 'great masses of the population' give their 'spontaneous' consent' to the 'general direction imposed on social life by a dominant fundamental group'. In this choice of terminology ('fundamental group' rather than 'class') we can see how Gramsci was deploying the concept of hegemony in an even more general sense than had been done within the discourse of the Third International, while at the same time, as has often been pointed out, attempting to divert the attention of the prison censors. But this relatively abstract notion of 'fundamental group' is also crucial to the appropriation of Gramsci's analysis by theorists of the new social movements, as we will see. At any rate, according to Gramsci, a group seeking 'supremacy' must 'lead' kindred and allied groups ('friends') that recognize and accept its moral, intellectual and political superiority. Here he assumes the existence of something like a liberal pluralism, that is, the existence of social actors who are capable of convincing, or being convinced, on the basis of reasoned argumentation.

It cannot be assumed that all social groups can be treated in this way, however. Thus there must come into play 'the apparatus of state coercive power which 'legally' enforces discipline on those who do not consent either actively or passively' (1971: 12). In times of 'crisis', Gramsci argues, a group seeking hegemony must strive to 'dominate' or 'liquidate' antagonistic groups ('enemies'), using armed force where necessary (57).[13] While he does not consistently specify the required order in which a dominant fundamental group must achieve both hegemony in civil society and state power in political society, it is clear that for Gramsci both are *necessary but not sufficient* conditions of a successful social transformation. No hegemony without state power; no state power without hegemony.

At other times, though, Gramsci analyses these two modes as though they exist in a hierarchical, rather than a complementary, relationship. In speaking of the 'two forms in which the State presents itself ... i.e. as civil society and as political society' (268), Gramsci elevates the state from its position *alongside* civil society to a position above both it *and* political society. In this model the state takes on the function of a dialectical completion or subsumption: it 'presents itself' as both civil and political society (263). Here the coercive apparatus of the state is given primacy over consensual processes, so that hegemony in civil society appears not as an end in itself, but rather as a means of achieving power by 'becoming' the state (261–3).

While it is clearly derived from liberal philosophy and nineteenth-century marxist theory, the privilege granted to the state in Gramsci's analysis is also driven by his empirical observations of modern revolutions, including the long struggle for the unification of Italy. Like most revolutionaries of his time, Gramsci was very interested in discovering how a social group of limited scope could achieve dominance over an 'entire national society' (1971: 56). He assumed that the natural and inevitable result of hegemony, as a pluralized play of antagonistic forces within the boundaries of a nation-state, was that 'only one' of the contending forces would 'tend to prevail ... to propagate itself throughout society' through its control of the state apparatus (181). That is, he assumed that the goal of any successful social transformation is to allow a group with a set of particular interests to bring about 'not only a unison of economic and political aims, but also intellectual and moral unity' (182). Only via the party, which Gramsci, following Machiavelli, calls the 'modern prince', can the germs of a collective will coagulate to become '*universal and total*' in this way (129; italics added).

Why are what Laclau and Mouffe (1985: 156) have called *irradiation effects* necessary? Gramsci's distinction between those whose consent should be sought and those who must be coerced is based on class (or group) self-interest—the boundary between consent and coercion is set, practically, by the relative distance between the demands of the group seeking hegemony and those of the groups to be led or coerced. Thus, civil society appears in fact as a political construction, a 'sphere' which does not pre-exist situated relations of power, but is created on the spur of the historical moment as it were, to separate those who might conceivably be brought into a revolutionary alliance from those who must be excluded from it.

GRAMSCI, LENIN AND THE HEGEMONY OF HEGEMONY

Thus I would suggest that Gramsci sought, as did Lenin, that revolutionary 'sweet spot' where dictatorship meets democracy. Debates abound, of course, regarding Gramsci's relation to Lenin and leninism, some of which are certainly relevant here. Femia argues that, both pre- and post-prison, 'the essential structure of [Gramsci's] thought and the core of his political commitment was marxist and revolutionary—albeit innovative and flexible' (1981: 243). The only point of gaining hegemony in civil society, Femia suggests, is to ensure the success of a full-frontal assault on state power. On this reading, which is also supported by Massimo Salvadori, 'Gramsci's theory of hegemony is the highest and most complex expression of leninism' (1979: 252). Norberto Bobbio, on the other hand, argues that 'Gramsci's theory introduces a profound innovation with respect to the whole marxist tradition. *Civil society in Gramsci does not belong to the structural moment, but to the superstructural one*' (1979: 30; italics in original). It seems to me that Gramsci's thought does indeed represent a break with certain aspects of Lenin's theory and practice, while at the same time continuing and deepening his rhetorical deployment of the play between dictatorship and democracy by extending it to liberal-capitalist societies.

This observation is made, of course, from a position relatively 'inside' the marxist tradition. From the point of view of the genealogy I am trying to construct, what is most important to note is that both Lenin and Gramsci theorize hegemony primarily as a mode of *political* revolution characteristic of what might be called the 'old social movements' (OSMs), that is, modern liberalism and marxism. As I have indicated in this chapter, political revolutionaries seek effects that (1) are to be felt over an entire social space, usually a nation-state, and (2) are expected to occur across a wide spectrum—indeed, the widest spectrum possible—of social, political, cultural, and economic structures and processes. Political revolutions are totalizing in their intent, and rely upon authoritarian, state-centred models of social change which, in the marxist tradition, are primarily or exclusively class-based and, in the liberal tradition, mask their class orientation by a reference to an illusory universality based on the forfeiture of individual autonomy. Despite their many differences, modern liberals and modern marxists share a fundamental assumption about the necessity of hegemony, an assumption that, as I will now show, is repeated on the terrain of postmodern societies.

3

Tracking the Hegemony of Hegemony: Postmarxism and the New Social Movements

> The term new social movements is rapidly approaching
> its sell-by date.
>
> (Crossley 2003: 149)

THE LONG MIDDLE OF THE TWENTIETH CENTURY

Before leaping into a discussion of postmodern politics, it is important to chart the declining years of western modernism. These were marked by the inability of the Russian, Chinese and other authoritarian socialist revolutions to realize the promises of marxist ideology and the successful co-optation of the western working class by the welfare state. Between the 1930s and the 1960s, liberalism was for a short period able to proclaim, at least in the 'developed' countries, that it had delivered on its promises, and in this (very small) portion of the world there was relative class peace for a while. However, it was not long before new forms of contestation emerged in response to old injustices within the nation-states of the Eurocolonial domain. The Black civil rights movement, second-wave feminism, environmentalism, insurgencies of indigenous peoples and the anti-nuclear movement challenged the liberal vision of a harmonious universality, declaring it once again to be an illusion. These so-called 'new social movements' (NSMs) shattered the precarious balance that had been achieved under the Keynesian accommodation, and established new streams of theory and activism that overturned many of the assumptions of the traditional marxist Left. Beyond Europe and its White-settler satellites, nationalist movements of liberation were continuing apace—more and more nations wanted *their own* place in the system of states. These struggles were not only nationalist, but often anti-capitalist in orientation, and they also made important contributions to radical theory and practice through the development of postcolonial critiques of Eurocentrism and racism. In the 1960s, it seemed that a brighter day had dawned for radical struggles all over the world.

But beginning quietly in the 1970s, a new force for social change began to take shape. It was reactionary, it was violent, but it was supported by many of the most powerful governments and corporations in the world. At this point it was largely clandestine, but by the 1980s it came to public awareness through the political figures who fronted for it: Ronald Reagan and Margaret Thatcher, the first 'neoconservatives' who pushed openly for a new world order based on the destruction of the Keynesian accommodation and a return to the laissez-faire doctrine of the most destructive phase of classical liberalism. Global institutions such as the World Bank and International Monetary Fund (IMF) began to be transformed via the so-called Uruguay Round of trade talks beginning in 1986. These brave new bureaucracies allowed neoliberal ideology increasingly to impose its policies upon countries of the global North and South, each in its own way, but always to the same effect: the rich White males got richer while the natural environment was increasingly depleted, as though to be sure that not only no one, but no thing, was left out of the cycle of capitalist exploitation.

At the end of the 1980s the Soviet Union collapsed, clearing the way for this nascent ideology, which came to be known by its opponents as neoliberalism, to increase its reach even further. This leaner, meaner version of the most successfully modern ideology is eminently suited to furthering capitalist globalization while at the same time appearing to overcome the exclusivity built into the modern system of nation-states. *Everyone* can be included in the Holy Communion of postmodern absolute capitalism, provided of course that everyone is willing to take up the position assigned to them by a racist, heterosexist, classist, ageist system hell-bent on its own destruction. Only now in the process of revealing itself, and still preferring secret talks among elites to open debate, neoliberalism represents a new incarnation of the logic of hegemony, one that combines the worst of marxist communist collectivism with the worst of liberal-capitalist individualism. Where liberalism was supposed to give us freedom, neoliberalism promises *security*, which is to be guaranteed by the integration of all human communities within a globalizing system of state-capitalist control. Thus does neoliberalism hope to attain a worldwide hegemony of a sort heretofore unknown in human history. An ambitious goal and an impossible one perhaps—

but one that certain political and corporate leaders clearly see as within their grasp.

Although much more can, and should, be said about the trends I have so briefly sketched out, my goal in this chapter will be restricted to showing how postmarxism has overcome some of the limits of western marxist theories of hegemony, but at the cost of a lapse into liberal pluralism. Given the abject failures of actually existing socialism, it is not at all surprising that some marxists tried to 'update' their approach. But in the context of the rise of neoliberalism, strategies based on representation, recognition and integration have become much more dangerous than they once were. These dangers are most apparent in the theory and practice of liberal multiculturalism, which attempts to extend the realm of harmonious plurality beyond the economy, to cover ethnic divisions and, in the limit case, to apply to *all possible* forms of difference. Despite this high goal, however, the so-called 'politics of recognition' has been rather poorly received by many of those whose concerns it is supposed to address. Theorists and activists from all over the world are making the links between this politics and the emerging neoliberal order. Not only are they making these links, but they are beginning to create alternative forms of autonomous identity and solidarity to contest them. This is, however, to anticipate too much. Before these alternatives can be situated in the history of radical social transformation, it will be necessary to address the next significant moment in the story of the logic of hegemony: the rise of the new social movements.

WHAT WAS NEW ABOUT THE 'NEW SOCIAL MOVEMENTS'?

Perhaps the best way to begin a discussion of the social movements of the 1960s is to ask what precisely it is that makes them 'new'. This is far from a simple question, since different analysts have produced different and mutually contradictory lists of defining characteristics, and disagreements on their applicability are rampant. As always, it should be remembered that social scientists create what they study in the process of studying it—that is to say that from a certain point of view, NSMs are an abstraction, a product of the sociological imagination and nothing more—or less—than that. There are, however, observable trends in the approaches of certain groups working in the western world in the 1960s–1980s, some of which I will now try to tease out.

Most new social movement theorists agree, for example, that NSMs differ from OSMs in addressing a wide range of antagonisms that cannot be reduced to class struggle—racism, patriarchy, the domination of nature, heterosexism, colonialism, and so on. This displacement of class as a fundamental antagonism has led many commentators to see NSM politics as 'merely symbolic' and individualistic (Melucci 1989: 5; Touraine 1992: 373; Pulido 1998: 7–8). Paul Bagguley uses the term 'expressive politics' to describe the activities of those he sees as 'bearers of a new hedonistic culture' of 'personal freedom' (1992: 34). Although this interpretation is widely accepted, it has a certain dismissive quality that I think needs to be challenged. Certainly, there are some individuals in some movements who relate to their activism on a purely personal level. But it is difficult to understand how striving to improve the situation of queers, women and people of colour, or working against military and ecological destruction, can be seen as individualistic pursuits. The burnout rate of activists in these movements would also seem to suggest that their struggles are no more pleasurable than those associated with class warfare. Hence, I would argue that the most accurate description of NSMs is not that they have *no* analysis of socially structured antagonisms, but that they do not focus solely on class as *the fundamental* axis of oppression. These struggles appear 'merely symbolic' or 'merely cultural' only in the eyes of those for whom economic concerns are the only important concerns, and who do not perceive the ways in which identity-based issues are intertwined with economic issues.

Another common observation is that NSMs are unlike their precursors in that they are 'not perceived to be struggling for a grand or universal transformation' (Pulido 1998: 8). For this reason, they are often cast as single-issue movements. Once again, while there is certainly some value in this observation, it is somewhat reductive and ignores long-standing analyses of relations *between* various struggles. As early as the 1970s socialist feminists were discussing links between patriarchy and capitalism (Firestone 1970; Eisenstein 1979), environmentalists were linking capitalism to the domination of nature (Leiss 1972; Bahro 1986), and so on. For these reasons, I do not accept without qualification the characterization of NSMs as single-issue struggles, and I would also want to challenge the dichotomy between liberal reform and marxist revolution that underlies this thesis. However, I would agree that operating across one or a few axes of oppression is a very different thing from seeking the wholesale reconstruction of an existing order through revolutionary means, and I take this shift as a marker of a partial undermining of the logic of hegemony.

This trend is also visible in the orientation of NSMs to state power. As the micropolitical, capillary nature of macrostructures and processes of power came to be more widely acknowledged, activist attention began to shift to a 'politics of everyday life and individual transformation' (Melucci 1989: 5). Also, and very importantly for the genealogy of the logic of affinity, the social movements emerging in the 1960s reflected a commitment to the notion that the means of radical social change must be consistent with its ends (Offe 1985: 829–31; Melucci 1989: 5; Bagguley 1992: 31). However, the absence of a totalizing conception of change and the recognition of the deep entwining of the personal and the political do not necessarily lead to a critique of state power as such. As many commentators have pointed out, the NSMs are characterized primarily by a politics of protest and reform (Bagguley 1992: 32; Touraine 1992: 392–3). The movements that are most commonly cited as exemplars of their type tend to desire irradiation effects across an entire social space, usually delimited as a national territory, and the changes most often cited as their successes have involved modifications to juridical structures. In NSM politics, then, there remains a strong orientation to the state, and this is a crucial moment of commonality between them and the OSMs they are usually thought to have superseded. The key difference between the two approaches, from the point of view that I am trying to develop, is that the NSMs hope to achieve effects on a *limited number* of axes, rather than on all axes at once. Thus I would argue that the dominant stream of the new social movements remains within a hegemonic conception of the political, and is only marginally and nascently aware of the possibilities inherent in actions oriented neither to achieving state power nor to ameliorating its effects. It is in this sense that they are correctly seen as a New *Left*, that is, as post*marxist* forms of struggle.

HEGEMONY GOES POSTSTRUCTURALIST: LACLAU AND MOUFFE

For my purposes, the most important theoretical development at this time was the reworking of Gramsci's concept of hegemony by a new generation of theorists who were steeped in Lacanian psychoanalysis and Derridean deconstruction. One highly influential product of this effort was Ernesto Laclau and Chantal Mouffe's *Hegemony and Socialist Strategy* (1985), which pushed Gramsci's theory to its limits in an attempt to understand and provide guidance to these emergent forces for social change. This book has been much maligned, uniting otherwise disparate factions in their distaste for its high style. Sharon Smith, a member of the International Socialists (IS), speaks for many when she observes that 'Laclau and Mouffe clearly share the postmodernist conviction that obscurity, abstraction and self importance amount to political sophistication, or at least create the illusion thereof. *Hegemony and Socialist Strategy* is therefore filled with run on sentences laden with jargon incomprehensible to those not already schooled in the language of postmodernism' (1994). While I can appreciate the frustration of those not familiar with the concepts that drive Laclau and Mouffe's work, it does seem unfair to attempt to judge the level of sophistication of a text that one cannot understand. At any rate, *Hegemony and Socialist Strategy* has had a major influence on how the concept of hegemony has been deployed within academic disciplines across the humanities and social sciences, and thus merits close and careful engagement. This can—and should—be done, using language and examples that are more accessible to non-specialist readers.

How, then, is this book relevant to the genealogy of the concept of hegemony? While celebrating the fact that 'in Gramsci, politics is finally conceived as articulation' (1985: 85), Laclau and Mouffe object to Gramsci's assumption that 'there must always be a *single* unifying principle in every hegemonic formation, and this can only be a fundamental class' (69; italics in original). Let us unpack this sentence. The conception of politics as 'articulation' means that particular historical tasks cannot simply be assigned to particular classes, as was attempted by Marx, Plekhanov or Lenin. Rather, identities (including but not limited to class identities) are formed through the growing realization of a common situation, and through struggle to improve this situation; their tasks are set not according to the playing out of a destiny, but by complex relations of social, political and economic power. Thinking of politics as articulation, therefore, has the effect of removing some of the more mechanical and authoritarian elements from marxist theory—no longer can anyone claim that his or her chosen strategy *must* be correct because it is 'in line with history'. The second half of this sentence is less obscure. In their anti-essentialist reworking of the theory of hegemony, Laclau and Mouffe displace not only the working class, but class as such, from the centre of radical struggles. Instead, they see class as one of many struggles that form a broad and indeterminate 'project for radical democracy'.

Given the previous discussion of the attributes of the new social movements, it should be clear that Laclau and Mouffe's efforts were directed towards tailoring the marxist theory of hegemony to respond to current trends in social activism. They explicitly link the project for radical democracy to characteristic struggles of the 1960s to 1980s, including the peace movement, as well as 'older struggles such as those of women or ethnic minorities' (165). But this list is not complete, and is indeed impossible to complete, since new sites of antagonism (potential political struggle) are constantly emerging, 'questioning the different relations of subordination ... and demanding ... new rights' (165). Laclau and Mouffe place this ongoing expansion of the field of rights in the context of what they call 'the democratic revolution', that is, the liberalizing upheavals that were discussed in the previous chapter, continuing on through the women's suffrage movements and the struggle for Black civil rights in the USA. Their basic thesis is that western societies are moving towards an increasing degree of democracy via an expanding notion of who should have access to liberty, equality and community. The project for radical democracy aims to push this process along, to accelerate and deepen it to the greatest extent possible, driven by an orientation to an imaginary end point which Jacques Derrida has referred to as a 'democracy to come' (1994: 59).[1]

Many marxist critics have questioned whether this project is indeed radical, given its abandonment of the centrality of class struggle and its adherence to what appear to be bourgeois values (Geras 1987; Bertram 1995). I want to raise a similar question, but on a different basis. I want to ask whether Laclau and Mouffe's theory takes us *far enough away* from classical marxism and the old social movements, far enough away from irradiation effects and the orientation to state power, to remain applicable in the context of the emerging struggles of the 1990s and 2000s. To this end I will discuss the exposition of the theory of hegemony found in Ernesto Laclau's contributions to a more recent text, *Contingency, Hegemony, Universality* (Butler, Laclau and Žižek 2000).[2] Despite the difficulties presented by its terminology and tight logical style, Laclau's analysis brings out something that is obscured in most of marxist and liberal theory—he allows us to see, with extreme clarity, the operation of the logic of hegemony.

In these essays, Laclau argues that there are four interlocking 'dimensions' of hegemony. First, he states that 'unevenness of power is constitutive of the hegemonic relation' (Butler, Laclau and Žižek 2000: 54). This is to say that hegemony occupies a middle ground between the war of each against each, where power is widely and evenly distributed, and the totalitarian regime, where all individuals and groups are subordinated to an overarching apparatus of control. The logic of hegemony, therefore, operates only in societies where there is a 'plurality of particularistic groups and demands' (55), that is, in *liberal* societies. In one sense, the statement of the first dimension of hegemony can be seen as a mere acknowledgement that something like a (post)modern condition exists within the liberal-capitalist world.[3] That is, it simply points out that politics today occurs on a complex terrain of relations within and between particular identities, corporations, states and groups of states. But, as I will argue later, there is also a normative component to the first dimension of hegemony, in the assumption that today's liberal societies represent the best, or perhaps the *only possible* mode of social organization that acknowledges and thrives upon this condition of unevenness of power.

The second dimension of hegemony holds that 'there is hegemony only if the dichotomy universality/particularity is superseded' (56). For Laclau, no political struggle can include everyone, since it is impossible for those who advance a cause to completely leave behind their own interests. Similarly, there is no such thing as a merely particular struggle, since no identity can exist without being in relationships with other identities, with what it is *not* (the 'constitutive outside').[4] In a 'hegemonic articulation', Laclau says, particular interests 'assume a function of universal representation', leading to a mutual 'contamination' of the universal and the particular (56). This process operates via the establishment of 'chains of equivalence', extended systems of relationships through which identities compete and co-operate, each seeking to enlarge itself to the point of being able to represent all of the others. Although the language deployed is different, Laclau is in fact following the lead of Gramsci in further generalizing the theory of hegemony put forward by Axelrod and Plekhanov. Establishing chains of equivalence means something quite similar to what Gramsci called 'leading' other classes, or what Marx called taking up one's historical role as the 'only truly universal class'. The difference is that, where Gramsci was able/compelled to speak of a 'fundamental group' rather than a class, Laclau speaks of an identity or, more precisely, an identification.[5] Marxists from Marx to Gramsci *thought they knew precisely* what was to be done and who had to do it. Laclau and Mouffe have an intimation of what is to be done (expand the democratic revolution), but refuse to say precisely how this is to be achieved or who is to achieve it. All of that, they argue, must be left to social actors themselves.

It is crucial to note that while the leading identity in a hegemonic articulation is itself part of the chain of equivalences, it simultaneously sets itself above it, via the elevation of its particular concerns to universal status (302). To the extent that the Green movement has been successful in its programme, for example, a diverse array of social groups have lined up under the banner of 'ecological sustainability', each expressing its own particular concerns about environmental destruction: parents as guardians of the well-being of young children; people of colour as those affected by environmental racism; and so on. This elevation has an effect not only on the hegemonized identities, however, but also upon the hegemonizing identity itself. As a corollary of the contamination of the universal and the particular, Laclau argues that hegemony 'requires the production of tendentially empty signifiers' which articulate chains of equivalence (207). The empty signifier—not to be confused with Lacan's floating signifier[6]— has a dual aspect. Empty signifiers are signifiers to the extent that they resonate within existing discourses; they do participate in the production of meaning. But they *tend* towards emptiness, or lack of meaning, due to the stresses placed upon them by their usage in a hegemonic articulation. That is, in order to be seen as a general equivalent for an increasing number of struggles, they must be ever further removed from their point of origin in a particular struggle. As an excellent example of an empty signifier, the term ' Green' will again suffice. It manages, with apparent ease, to refer to mainstream political groupings oriented to parliamentary reform (Green Party), underground movements that carry out direct action against the destruction of the environment and in defence of non-human beings (Green Warriors), and niche-marketed products in the capitalist marketplace (Green Detergent). The result of all of this overtime is that most of us are not at all sure what it means to 'be Green'. This signifier tends to emptiness, or lack of meaning, precisely because of its fullness, its multiplicity of meanings.

Finally, Laclau argues that '[t]he terrain in which hegemony expands is that of a generalization of the relations of representation as condition of the constitution of the social order' (207). With this thesis, we appear to have returned to the empirical realm of the first dimension; under conditions of (post)modernity, representation—or the delegation of power in the economy, cultural production and political will formation—becomes 'the only way in which universality is achievable' (212). However, once again we must be aware that this is no mere description. The claim being made is not only that mass representation is necessary, but that it is *desirable*, because it is through processes of representation that equivalential chains are expanded, hegemonic blocs are formed and social transformations are achieved. This theoretical argument has been taken up in analyses of many popular struggles, such as those for and against Thatcherism in the UK (Hall 1983), Reagan–Bush conservatism in the United States (Grossberg 1992: 377–84), and studies of the role of television in maintaining consent to the established order of racist, sexist, capitalism (Kellner 1990; Press 1991). The strength of these analyses is that they move beyond the Frankfurt School's postulation of a one-dimensional apparatus of ideological domination, in which possibilities for resistance are negligible or non-existent. Their weakness is that, in valuing contestation as such, they do not always pay enough attention to the precise logic of various *modes* of contestation or acknowledge that a diversity of logics of struggle exists. More precisely, they tend to advocate only for *counter*-hegemonic struggles against various kinds of subordination. Lawrence Grossberg's 'affective politics', for example, sees the struggle for hegemony as a 'struggle for authority' (Grossberg 1992: 380–1). And Douglas Kellner echoes Laclau's thesis on representation quite closely in claiming that '[b]ecause of the power of the media in the established society, any counter-hegemonic project whatsoever—be it that of socialism, radical democracy, or feminism—must establish a media politics' (Kellner 1990: 18). Thus the theory of the new social movements takes off from the pseudo-democratic dictatorship of Lenin and Gramsci only to land on the dictatorial pseudo-democracy of Hobbes and Locke.

In suggesting that postmarxism's line of flight from communism takes it into liberal-capitalist territory, I do not mean to imply that these two ideologies are equivalent. They differ in the radicality of their critique of most existing institutions, for example, and are directly opposed in that postmarxists tend to favour a socialist rather than a capitalist liberal democracy (Bobbio 1987; Mouffe 1993: 90–101; Laclau 1996: 121). My point is rather that postmarxism and liberalism rely upon a similar *logic*, a logic of representation of interests within a state-regulated system of hegemonic struggles. The expected outcome of the representation of a situation of inequality or lack of rights is *recognition* of the oppressed identity by the state apparatus. Recognition is supposed to lead to an improved situation for the identity in question through its inclusion in the list of 'those who are to be granted equal rights'; that is, through its *integration* into the hegemonic social order. Oft-cited examples of the success of this model are women's suffrage, the gains made by the struggle for Black civil rights in the US, and the various multiculturalism policies adopted by countries like Canada and Australia. In order to show more precisely how postmarxism and contemporary liberalism share a common logic, I will now spend some time discussing the concepts of recognition and integration as they have been developed in the theory and practice of multiculturalism.

LIBERAL MULTICULTURALISM AND
THE RECOGNITION/INTEGRATION PARADIGM

The term multiculturalism is notoriously slippery, evoking shades of meaning that are particular to certain national and international contexts, and which are changing rapidly within these contexts.[7] The discussion to follow will attend to this diversity of meanings, will in fact *rely* upon it as links are made between liberal multiculturalism and neoliberal globalization. But I will begin by addressing liberal multiculturalism as state policy in Canada, Australia and the European Union.[8] This style of politics is often supported by Charles Taylor's argument, in 'The Politics of Recognition' (1992), that it is crucial for established state peoples to 'recognize' those identities that have been historically excluded from full citizenship rights, that 'we'—the current state peoples—must engage 'them'—the historically excluded—in a 'dialogue'.

To understand the sense in which Taylor uses the terms recognition and dialogue we must delve into the related notions of identity and authenticity. Taylor bases his theory of identity on a Herderian ideal of authenticity as 'self-realization'. In this modernist conception of the self, which directly influenced Hegel, both the individual within a particular culture and the culture itself must strive to be 'true to' their inner nature. Since the individual can only achieve self-expression within a culture, Taylor argues, 'we define our identity always in dialogue with, sometimes in struggle against, the things our significant others want to see in us' (1992: 30–3). What we are searching for in this process, Taylor argues, is 'recognition' or 'acceptance of ourselves by others in our identity' (1993: 190). Now, while Taylor sees the need for recognition within a culture as 'a crucial feature of the human condition', he suggests that modern individuals are unique in confronting the possibility that recognition might *fail* (1992: 32). Thus a politics of recognition has arisen, a series of demands for acceptance of previously marginalized identities. For Taylor, the appropriate response to these demands, at least with respect to cultural identities, is to seek out what he calls 'as yet unexplored modes of deep diversity', through which particular nation-states would acknowledge the existence of a multiplicity of nationalities and ethnicities within their borders (1993: 200). This acknowledgement, he hopes, will provide a basis for what he has called a 'post-industrial *Sittlichkeit*' (1975: 461)—a merging with a Spirit, to be sure, but with a Spirit of unity within diversity.

Liberal theorist Will Kymlicka's conception of 'differentiated citizenship rights', which draws on the work of Iris Marion Young (1989), can be seen as an extension and practical-theoretical elaboration of Taylor's argument. Like Taylor, and following Hegel, Kymlicka assumes 'different' cultures must express themselves 'differently' within the system of nation-states, and be recognized according to their authentic nature. Each of the three categories he considers as important to the context of contemporary state formations—colonizer, national minority and immigrant—is assigned a varying degree of sovereignty, through an evaluation of the validity of its claim to possess what he calls a 'societal culture'.[9] Colonizers, or majorities, Kymlicka argues, can more or less look after themselves, as they tend to have already achieved full citizenship rights. National minorities—those who have been colonized—should be granted limited forms of self-government, such as has been the case with the Basques in Spain, Quebecois in Canada and Aboriginal peoples throughout the Eurocolonial domain. Immigrant identities, as neither colonizing nor colonized, are not able to claim support for their own societal cultures, this right being 'neither desirable nor feasible' for them (1998: 35). Rather, they are expected to 'integrate' with the dominant society. Like Taylor, Kymlicka argues that his system of multicultural citizenship, in the context of an asymmetrical multinational federation, offers the best, if not the only, model that will allow states facing the pressures of the politics of recognition to survive.

While the multinational federalism that Kymlicka and Taylor advocate is clearly an advance over the modern Hegelian nation-state, I'm not convinced that it is adequate to the demands of the postmodern condition. As a way of exploring the deficiencies of this model, we might compare Taylor's notion—and performance—of dialogue with the resonances given this term by Bakhtin and Hegel, whom Taylor explicitly cites as precursors (1992: 26, n. 13; 33 n. 9; 34 n. 10). Central to the Bakhtinian notion of dialogue is the absence of any finalizing, totalizing or all-knowing position—the absence, in fact, of a hegemonic moment.[10] In Taylor's texts, however, we find claims like the following: 'The demand for recognition tends to hide itself, tends to be presented as something else' (1993: 192). Such a claim cannot be made without invoking a privileged field of vision that is capable of assessing the 'true' intentions that lie behind the statements of others. Here Taylor is not playing the game as he says it should be played; he is not taking these identities for 'what they really are', or at least for what they *say* they really are. A similar moment occurs in Kymlicka's argument when he *assigns* certain political motivations and rights to *all* who are supposed to occupy each of the citizenship categories he creates. Here again, we find an utterance that does not anticipate a rejoinder. The speaker already knows both what his addressees 'really want', *and* how to give it to them.

In keeping with its anti-totalizing character, Bakhtin's polyphonic notion of dialogue is predicated upon an assumption of radical equality. It presupposes a 'plurality of consciousnesses of *equal value*, together with their worlds', a plurality of 'independent and unmerged voices' (Bakhtin 1984: 4–5). Taylor's conception of dialogue is again clearly not Bakhtinian, as we can see in his inability to accept the demand that 'we all *recognize* the equal value of different cultures; that we not only let them survive, but acknowledge their *worth*' (Taylor 1992: 64). The problem for Taylor is that 'mere difference can't itself be the ground of equal value' (1991: 51). Rather, the judgement of equal worth that leads to recognition is for him an *empirical* question. He writes: 'On examination, either we will find something of great value in [a given] culture ... or we will not. But it makes no more sense to demand that we do so than it does to demand that we find the earth round or flat, the temperature of the air hot or cold' (69). That is, he continues, 'if the judgment of equal value is to register something independent of our own wills and desires, *it cannot be dictated by a principle of ethics*' (69). Here Taylor leaves both Bakhtin and Hegel far behind, and again seems to be working against his own position, by presenting a theory of ethical community that precludes ethical considerations.

Taylor's form of recognition not only contains monological and anti-egalitarian elements, it is also based on what, in Hegelian terms, could only be considered a 'deficient' mode of recognition associated with the master–slave relationship: a recognition that Hegel describes in the *Phenomenology* as 'one-sided and unequal', since one partner (the master) is 'only recognized', and the other (the slave) is 'only recognizing' (Hegel 1977/1807: 113, 116). The key here is that recognition is not mutual or, as Bakhtin puts it, there is no 'affirmation of the other's consciousness as a full-fledged subject' (1984: 7). Shifting the emphasis in the quote from Taylor cited above, we can see how a one-way motion takes place: 'that *we* not only recognize *them*, but acknowledge *their* equal worth'. Although he has recently castigated those who are 'still so used to functioning politically among themselves' that they continue to 'speak, think, and act politically in terms of us and them' (1998: 146) it would seem that Taylor's own theory of recognition is deeply immersed in this mode. Recognition is something that 'we' may or may not wish to bestow upon 'them', depending upon whether 'we' judge 'their' particular claim to be valid. Again, a similar operation is carried out by Kymlicka: 'I will discuss whether immigrant groups *should be given* the rights and resources necessary to sustain a distinct societal culture' (1995: 76; italics added). Or: 'If people have a deep bond with their own culture ... *should we not allow* immigrants to re-create their own societal cultures?' (95; italics added).

On the theory that individuals attempt to express their true identities in dialogue with others, it would be fair to ask how statements such as these might be motivated. That is, what is the nature of the 'objective' point of view from which a system of empirically-guided recognition and differentiated citizenship rights could be formulated? This question is easy to answer in the case of Taylor, who clearly relies upon a model of the nation-state which he associates with 'North Atlantic civilization' (1992: 71). But with Kymlicka it's not so easy, since he doesn't position his theory in this way. He does, however, argue that 'the crucial question facing ... any ... multination state is how to reconcile ... competing nationalisms within a single state' (1998: 127). He also has a tendency to cast any non-universal identity-building project as a 'minority nationalism' giving rise to 'disintegrating effects' (132). Kymlicka's point of identification, it seems, is not with a particular ethnic group, category of citizenship, or even with an elusive, non-exclusive nationalism, but with *the system of states itself*. From this 'privileged empty point of universality' (Žižek 1997: 44) that presumes to stand outside the realm of ethnocultural identification, all other positions can be categorized according to their presumed eligibility for, and willingness to accept, a certain model of multinational federalism that preserves the hegemony of the existing social, political and economic institutions.

What is common to the theories of Taylor and Kymlicka—and to multiculturalism as state policy in general—is its choice of Hegelian dialectics over Bakhtinian dialogism. Of particular concern here is the role attributed to the state within the dialectic of recognition. On this point Axel Honneth has presented a very interesting argument, which I would like to quote at length. Honneth suggests that, because Hegel ultimately chooses spirit over intersubjectivity in his account of recognition:

> the construction of the ethical sphere occurs as a process in which all elements of social life are transformed into components of an overarching State .... In the State, the universal will is to collect itself into a unity, into the point of a single instance of power that must, in turn, relate to its bearers ... the way it relates to figures of its Spiritual production. Therefore, Hegel can do nothing but depict the sphere of ethical life on the basis of the positive relationship that socialized subjects have, *not among each other, but rather with the State* (1995: 58; italics added)

POLITICS OF DEMAND/ETHICS OF DESIRE

It is at this point, I would suggest, that liberal multiculturalism runs into a dead end. It assumes the existence of the state as a neutral arbiter, a monological consciousness that, upon request, dispenses rights and privileges in the form of a gift. This is, of course, precisely the same assumption made by postmarxist theories of representation and radical democracy—they also rely upon the hope that 'we' will be able to compel/persuade state and corporate apparatuses and other social structures to give 'us' (a little more of) what 'we' think we need. Both liberalism and postmarxism, then, share a reliance upon a politics of demand, a politics oriented to improving existing institutions and everyday experiences by appealing to the benevolence of hegemonic forces and/or by altering the relations between these forces. But, as recent history has shown, these alterations never quite produce the kinds of 'emancipation effects' their proponents expect. The gains that are made (for some) only appear as such within the logic of the existing order, and often come at a high cost for others. One well-known example is the 'integration' of women into the paid workforce of the middle and upper classes of the global North. Women who had been locked into the 'private sphere' of the home were set free only to be subjected to corporate domination and exploitation. The men stayed at work, of course, and children did not become suddenly able to look after themselves, so for every upper-class white woman who has been 'freed', several lower-class women (usually women of colour) have been 'enslaved' as child-care workers, nannies and housekeepers.

This spiral is particularly worrisome as nation-state-based liberalism is transforming itself into a global neoliberal order. That a discussion of multiculturalism as state policy should veer into a critique of a multiplicity of global relations of power is no coincidence. Recognition and integration are, and have always been, about much more than culture and ethnicity. Although it is almost unfair to choose such an easy target, the work of Francis Fukuyama provides a sense of the highest, wildest imaginings of neoliberal multiculturalism. Fukuyama claims that '[t]here is a fundamental process at work' in the late twentieth and early twenty-first centuries, a process that 'dictates a common evolutionary pattern for all human societies—in short, something like a Universal History of mankind in the direction of liberal democracy' (Fukuyama 1992: 48). Citing the collapse of the Soviet Union and the Chinese turn to capitalist markets, Fukuyama suggests that people everywhere are accepting liberal values of the sanctity of private property and enterprise, and heeding the call of a theological-historical force ushering in a new millennium based on 'discoveries about the nature of man as man' (51). Fukuyama goes so far as to suggest that not only is liberal capitalism the best guarantor of individual freedom in theory, but is in fact delivering this freedom in practice. In his bizarre world, the global South is 'rapidly closing the gap' with the North (41), so that while underdevelopment theory was once dominant in Latin America and Africa, these countries have now come on board with the development paradigm. With Fukuyama, Hegel is stood back on his head, and liberalism appears once again as the final, steady state of human history, in which all antagonisms have disappeared and the individual is raised up into (integrated with) the Spirit of Absolute Capitalism.

However, just as Taylor and Kymlicka have been forced to ignore their anti-integrationist critics so as to be able to continue with a simulated and carefully controlled 'dialogue', Fukuyama does not like to admit that the economic and political elites who have signed on to 'structural adjustment programmes' are an extreme minority, one that in many cases has been coerced into the neoliberal fold at the barrel of an economic gun. He also tends to ignore the eruption of struggles, all over the world, *against* his vision of the endtimes. The website of People's Global Action lists 33 'heavily indebted poor countries' that are involved in resistance to various aspects of neoliberal globalization (People's Global Action 2003). This list is long, and impossible to represent adequately in any summary. But to show how utterly blind the advocates of (neo)liberal triumphalism are to the horrid underside of their vision, I think it is worth describing at least some of its manifestations. In Bangladesh, garment workers who produce 'cheap' clothing for consumption in the global North, and who are mostly women, are struggling to free themselves from oppressive and dangerous working conditions, including compulsory overtime, rape and inadequate or non-existent support for child-care and maternity leave. In Ghana, a World Bank-funded water privatization programme threatens to turn over control of the most precious resource there is to corporate interests. According to Amenga-Etego of the non-governmental organization Integrated Social Development Centre (ISODEC):

> Where cost-recovery becomes the underlying policy, water will become unaffordable for many poor people in Ghana. Even before the project kicks off, taps are being turned off because a growing number of families cannot afford to pay. (Mutume 2000)

In South Korea, workers opposing the sale of the assets of the state power utility, Korea Electric Power Corp, to local and foreign corporate interests, are beaten and jailed by police (Seok 2000), and other groups, including students and farmers, are organizing to oppose the 'opening' of the Korean agricultural market. Although their situations and ideologies are disparate, and although they are not always in direct contact with one another, it is clear that many people all over the world explicitly reject the neoliberal dream of a global political economy united by a global multicultural identity.

The links between multiculturalism and neoliberalism have been interestingly analysed by Slavoj Žižek. He sees a progression in global political economy, from capitalism within sovereign nation states, with some trade, to colonialism, where one state dominates another, to the current situation, where there are no colonizing states, but rather all states are colonized by multinational corporations. While I would not quite agree with the thesis that the state form is in decline—a plethora of new supranational institutions is emerging not only to continue, but to intensify the discipline and control functions of the nation-states—I think Žižek is quite correct in his analysis of how multiculturalism and neoliberalism work together. 'The ideal form of ideology of this global capitalism', Žižek notes, 'is multiculturalism, the attitude which, from a kind of empty global position treats each local culture the way the colonizer treats colonized people—as "natives" whose mores are to be carefully studied and "respected" (1997: 43–4). In taking up this 'privileged empty point of universality', I would argue, neoliberal capitalism goes where Charles Taylor fears to tread—it has no problem with judging other cultures or with finding them to be of 'worth'—quite literally, to be of 'value'—as niche markets for consumption, servile populations for consumption, or opportunities for fire-sale appropriation of natural and humanly constructed 'resources' via privatization of national 'assets'. As Žižek puts it, globalizing neoliberal capitalism shows us how to 'appreciate (and depreciate) properly other particular cultures—the multiculturalist respect for the Other's specificity is the very form of asserting one's own superiority' (44).

Indeed, pursuing a politics of demand in the context of neoliberal globalization is rather like pursuing the latest in automobiles, clothing or refrigerator styles. One feels a lack, which one hopes to fill, only to discover that the yearning for fulfilment has increased rather than decreased. Just as no product can ever provide satisfaction in the consumption of goods and services, no state-based system of representation can be an adequate substitute for the autonomous creation of a just life lived in community with human and non-human others. Neoliberalism plays on the liberal and postmarxist hope that the currently hegemonic formation will recognize the validity of the claims presented to it, and respond by producing effects of emancipation. Most of the time, however, it does not; instead it defers, dissuades or provides a partial solution to one problem that exacerbates several others. In order to 'free' some educated upper-class First World White women to participate in the paid workforce, liberal capitalism creates new categories of indentured labour designed to import and enslave women of colour from the global South. So that we might achieve equality in the possession of private vehicles and air conditioners, the air becomes unbreatheable and the (newly privatized) power grid collapses in the heat wave associated with global warming produced by ... cars and air conditioners.

The irrational behaviour associated with the politics of demand can be understood only on its own terms, that is, on the terms of unconscious desire. In Lacanian theory, desire does not compel us to achieve satisfaction of a need. Rather, it compels us to continually *avoid* achieving satisfaction, since doing so would mean the end of desire and, in a sense, the end of the one who desires. It is for this reason that I say that postmarxist and liberal politics are driven by a *fantasy* of emancipation within existing structures of domination. 'What the fantasy stages', Žižek notes, 'is not a scene in which our desire is fulfilled, fully satisfied, but on the contrary, a scene that realizes, stages, the desire as such' (1991: 6). The role of fantasy is to 'give co-ordinates to the subject's desire', to place us in situations where we can *continue* to desire. Lacanian theory also helps us to understand why those who identify with the (neo)liberal and (post)marxist projects are resistant to any talk of operating non-hegemonically. 'Anxiety occurs not when the object-cause of desire is lacking ... . On the contrary, the danger is in our getting too close to the object and thus losing the lack itself. Anxiety is brought on by the disappearance of desire' (Žižek 1991: 8). Because they share an unconscious desire to perpetuate the desire for emancipation by extra-individual, extra-community structures of coercive power, (neo)liberalism and (post)marxism can be said to participate in an *ethics of desire*, a set of principles and outlooks that perpetuate a self-imposed failure and provide a cover for the abdication of the difficult tasks associated with autonomous individual and communal self-determination.

Breaking out of this trap is not at all a simple or easy process, although some political subjects have begun to do it—hesitantly, partially, implicitly. The effects of the failure of the recognition paradigm are visible in a large number of increasingly linked theoretical paradigms and activist strategies, which are renouncing the desire for representation, recognition and integration within the currently hegemonic order. The resurgence of direct-action strategies and tactics discussed in Chapter 1 is a manifestation of this trend in activist practices. I will now turn to a discussion of how certain theoretical paradigms are also pushing beyond the politics of demand and the ethics of desire.

'WE' ARE NOT 'YOU': ANTI-INTEGRATION THEMES IN POSTCOLONIAL, FEMINIST AND QUEER THEORY

As the policy of a particular state or group of states, liberal multiculturalism is very much a result of the historical expansion, and the more recent implosion, of European colonialism. According to the discourse of modern imperialism, 'the problem of difference' began 'out there', as 'they' always seemed to refuse the gifts of European civilization; but it has now progressed to the very core of western liberalism itself, where the same thing seems to be happening. That is, if the societies of the global North are indeed multicultural in the sense that those who identify as descendants of European nationalities are in many cases no longer a numerical majority, if there is indeed a multiplicity of cultural heritages at work within these societies, then how can liberalism posit itself as a universal medium for these societies? As Bhikhu Parekh has recently pointed out:

> To call contemporary western society liberal is not only to homogenize and oversimplify it but also to give liberals a moral and cultural monopoly of it and treat the rest as illegitimate and troublesome intruders. (2000: 112)

Homi Bhabha has also expressed his suspicions regarding a 'multiculturalist pluralism that dreams of a federation of 'minority' or ethnic groups stitched together in a multi-culti quilt'. Such efforts 'aspire towards the assimilative' and 'neglect the problems of power differentials, conflicts of interest, and cultural dissonance' (Bhabha and Comaroff 2002: 17). Indeed, postcolonial theorists and activists have been struggling since the late 1950s to bring to attention the many ways in which liberalism relies upon a deep-rooted Eurocentrism. 'The native is declared insensitive to ethics', wrote Frantz Fanon. 'He represents not only the absence of values, but also the negation of values' (1963: 41). To shift our attention from how Europe has imagined Africa, but to continue with the exposition of the Eurocolonial worldview, Edward Said has pointed out how 'from earliest times in Europe the Orient was something more than what was empirically known about it' (1978: 55). Like Darkest Africa, Exotic Asia was thought to be utterly thronging with marvels and curiosities of all sorts. Although it claimed to have cornered the market on Enlightenment, Europe's ignorance also served it well— people without values, people hardly human, could be treated in ways that would otherwise be unthinkable.

The construction of a fantasy space in which the exotic Other might be housed is of course not limited to the period of modern colonialism. As Hawaiian indigenous rights activist Haunani-Kay Trask has pointed out, even today 'people think of the Pacific Basin as a fantasy place where you run away from the cold in the middle of a cold winter. Their image of the Pacific Basin is a highly fantasized, romanticized image that has more in common with a home movie than with the actual geographic and cultural place' (1993). Although the method of simultaneously romanticizing and subjugating 'problematic' identities remains the same, the means by which some national formations dominate others are changing. The case of Hawaii is interesting in this regard because it blurs the boundaries between 'overseas' and 'internal' colonization, and brings to light the intense contradictions that are generated by attempts to integrate indigenous populations into the system of states. Taylor (1993: 200) and Kymlicka (1998: 30–1) have both argued that Indigenous peoples—as 'national minorities'—have a valid claim not only to recognition of their cultures, but also to self-determination via various forms of 'self-government'. The problem is, there is no cross-cultural agreement—or even any discussion!—on what a 'self' is and what it means to be 'governed'. Theorists like Taylor and Kymlicka, and states like Canada and Australia, hope to address issues of indigenous self-determination at the 'territorial, regional, and community levels' (DIAND 1997: 14). That is, their vision of self-government involves an attempt 'to *delegate* parliamentary authority ... not to *substitute* [indigenous] authority for parliamentary authority' (Long, Little Bear and Boldt 1984: 73; italics in original).[11] On this model, self-governing communities would have some say in matters of society, economy, polity as well as culture, but they would remain under the ultimate control of the Eurocolonial state apparatus. Thus it has been suggested that this approach may, in its current form, serve primarily to assuage the anxieties of semi-peripheral capitalist nation-states rather than to advance the goals of indigenous peoples (Povinelli 1998).

Indeed, in the Canadian context, there is a profound doubt among indigenous peoples about the value of being incorporated into the national-multicultural context. Marianne Boelscher-Ignace and Ron Ignace argue that it 'homogenizes' Aboriginal peoples into a '"native slot" on the ethnic landscape ... rather than acknowledging Aboriginal nations' specificity and rights to express this specificity on our own terms' (1998: 150). The incommensurability between the European system of states and indigenous modes of self-determination has been clearly articulated by academics and activists working in the field of Native American political theory.[12] Through a creative revaluation of their own histories in contemporary contexts, writers such as Taiaiake Alfred (1999), Vine Deloria (Deloria and Lytle 1984), Lee Maracle (1996) and Patricia Monture-Angus (1999) are self-consciously walking a fine line between parochial forms of cultural essentialism and genocidal integration into European modernity. They have proposed alternative visions of relations between individuals, human communities and the natural environment, and they have challenged the inevitability and desirability of centralized bureaucracy, capitalism and the state form. Through this process of intense critique, they have confronted the liberal illusion that mere 'recognition' of 'cultural difference' can lead to harmonious coexistence. And they have helped to show how policies of state multiculturalism, as progressive as they might be, direct political and academic attention away from many more pressing concerns.

For example, even if some Aboriginal communities manage to avoid the worst effects of rational-bureaucratic domination and capitalist exploitation in their quest for self-government, there remains yet another 'gift' of Western liberalism that many are reluctant to accept: patriarchy. 'The denial of Native womanhood is the reduction of the whole people to a sub-human level', writes Lee Maracle. 'The dictates of patriarchy demand that beneath the Native male comes the Native female' (1996: 17). On this point at least liberal multiculturalism *has* had something to say: gender inequality is sometimes considered as an example of a 'failure of recognition' (Taylor 1992: 27). However, when gender is considered, the simple fact that multiculturalism is a *liberal* discourse militates against the appearance of 'loaded' concepts like patriarchy and oppression. A nod to an equality-based, or perhaps a 'differential equality-based' form of 'citizenship' is the most one can expect (Mouffe 1993; Young 1989). Looking beyond liberal feminism, however, we once again encounter many voices challenging the logic of recognition and integration. If we see the pursuit of equality with men as a quest for recognition, then radical feminism's rejection of patriarchal values, cultures and social movements can be seen as a challenge to the integration paradigm (Dworkin 1974, 1989; MacKinnon 1989).[13] Socialist feminism made the same gesture with respect to the places of women in capitalism (Smith 1977; Eisenstein 1979); anarcha-feminism adds to this a refusal to be co-opted by state apparatuses (Ackelsberg 1991; Kornegger 2002); and Black and postcolonial feminists have pointed out that the figure of 'woman' whom White liberal feminists want to liberate does not resonate with their own experiences (Lorde 1984; Mohanty 2003). This difficult—but very productive—trajectory led to the feminism/postmodernism debates of the 1990s, which engaged with the question of whether it is theoretically justifiable and politically effective to seek the emancipation of some generalized entity called 'woman' (Nicholson 1989; Benhabib 1995). I will discuss some of the details of these debates in Chapter 6. For now, I want only to point out that postmodern feminism has arrived at a very interesting point in the critique of the quest for recognition by asking whether it is possible to imagine a politics that seeks 'neither to liberate a female subject nor to secure certain fundamental rights for her' (Elam 1994: 77).

A very similar path has been followed by the gay and lesbian liberation movements, which began as liberal quests for acceptance and recognition guided by the metaphor of 'coming out of the closet'. If only gay men and lesbian women could be treated the same as straight men and women, the argument ran, then the inequalities associated with sexual orientation would diminish or cease to exist. But the queer critique of relations between hetero- and homosexuality has pointed out how these two discourses are mutually dependent upon one another. Just as hetero requires homo in order to know what it is (not), homo demands hetero as *its* other. Thus Judith Butler argues that 'the affirmation of homosexuality is itself an extension of a homophobic discourse' (1993a: 308). Rather than claiming our sexuality, she argues, it might be more effective to *disclaim* it, to refuse to answer the question about what we do with our bodies in order to achieve pleasure: 'I come out only to produce a new and different closet' (309). Butler notes, of course, that disclaiming non-heterosexual identities may appear to be just what the forces of neoconservatism would like to see. But disclaiming is not the same as silence. By coming out into *an open field*, rather than into a hierarchical structure of fixed identities, she suggests that it is possible to undermine the coercive regulation of sexuality as such. This observation can easily be extended to other identities, which once again brings us up against the limits of the politics of demand/ recognition/integration.

TOWARDS A POLITICS OF THE ACT

As 'pragmatic' as it may be, and despite its successes during the heyday of the welfare state in a few countries, the politics of demand is by necessity limited in scope: it can change the content of structures of domination and exploitation, but it cannot change their form. As Laclau points out, without a hegemonic centre articulated with apparatuses of discipline and control, there is no force to which demands might be addressed. But the converse is also true—every demand, in anticipating a response, *perpetuates* these structures, which exist precisely in anticipation of demands. This leads to a positive feedback loop, in which the ever-increasing depth and breadth of apparatuses of discipline and control create ever-new sites of antagonism, which produce new demands, thereby increasing the quantity and intensity of discipline and control.

It is at this point that a politics of the act is required. This politics can be productively understood in terms of what Lacan has called the *ethics of the real* (Lacan 1992). According to Žižek, the force of this ethic derives from 'going through the fantasy', from 'the distance we are obliged to assume towards our most "authentic" dreams, towards the myths that guarantee the very consistency of our symbolic universe' (Žižek 1994: 82). Clearly, the fundamental fantasy of the politics of demand is that the currently hegemonic formation will recognize the validity of the claim presented to it and respond in a way that produces an event of emancipation. Most of the time, however, it does not; instead, it defers, dissuades or provides a partial solution to one problem that exacerbates several others. Going through the fantasy in this case means giving up on the expectation of a non-dominating response from structures of domination; it means surprising both oneself—and the structure—by inventing responses that preclude the necessity of the demand and thereby break out of the loop. This, I would argue, is precisely what is being done by the affinity-based networks of radical activism I have discussed in Chapter 1, and what motivates the anti-integrationist elements of postcolonial, feminist and queer theories.

A final point needs to be made here. At one stage in the development of his thinking on these matters, Žižek argued that an ethic of the real would demand not only that we traverse our own fantasy, but also that we 'respect as much as possible the other's 'particular absolute,' the way he organizes his universe of meaning in a way absolutely particular to him' (1991: 156). While Žižek later tried to step away from what he came to see as a 'universalizing' moment in his own thought,[14] I would suggest that it is impossible to consider ethical and political questions without reference to others, however defined, since such others will always necessarily *be* defined—if there is no Absolute Subject, then every identity is differential. It is, of course, precisely this kind of argument that leads Žižek to back off from the social-political implications of the ethics of the real. But the impossibility of a purely universal identity does not relieve us from the necessity of attempting to be in solidarity with others—note that I say solidarity, not identity. Solidarity occurs *across* identifications, which means that without a multiplicity of subject positions there can only be identity of struggles, at which point the concept of solidarity becomes meaningless. Thus I would see in the disavowed underside of Žižek's version of the ethics of the real a necessary element of the politics of affinity. Affinity-based action *is* ethical action, though it is clearly not moral, that is, not universalizing or totalizing in intent.

This distinction is extremely difficult to comprehend from within the liberal and marxist traditions, with their common basis in Hegelian notions of totality and ethical community. Morality, ethics and universality are seen within these paradigms—including their 'postmodern' variants—as inextricably bound to one another, so that it is impossible to conceive of an ethical act that is not based on a moment of universalization, or at least universalizability. If we want to understand how non-universalizing modes of social organization can be ethical, we need to trace their descent through lines of anarchist, rather than marxist or liberal, theory and practice. That is, we need to produce a *genealogy of affinity* that would show how this non-universalizing ethic is ever-present, but submerged under the hegemony of hegemony. This will be the task of the following two chapters, which will chart the difficult development of the logic of affinity, from eighteenth- and nineteenth-century anarchisms through to poststructuralist, autonomous marxist and, most recently, postanarchist interventions into the field of radical political theory.

4

Utopian Socialism Then ...

GUIDING THREADS

As is always the case with a genealogical approach, abandoning the search for origins leaves one facing a potentially infinite field. When and where to begin what has no beginning? Thinking within the western tradition, there is no doubt that pre-Socratic thought would be an important site for exploring the logic of affinity, as would the Gnostics, Rousseau, Blake, Goethe, Wilde ... the list goes on. Looking beyond the West, Sufism, Taoism and Zen, as Hakim Bey and others have suggested, offer tantalizing glimpses of the possibilities of a non-totalizing philosophy and practice. However, in keeping with my interest in how affinity and hegemony have been circulating within the (neo)liberal system of states, I will limit the discussion in this chapter to the rise of modern western socialism.

This is itself, of course, an extremely wide field. Within it, I will focus on three guiding threads. One of these is the distinction between social and political revolution, which develops out of an anarchist theory of social change that challenges, and ultimately breaks down, the dichotomy between revolution and reform. In *Paths in Utopia*, Martin Buber presents a genealogy of 'the utopian element in socialism' that tracks the emergence of what he calls 'structural renewal':

> [I]n 'utopian' socialism there is an organically constructive and organically purposive or planning element which aims at a re-structuring of society, and moreover not at one that shall come to fruition in an indefinite future after the 'withering away' of the proletarian dictator-state, but beginning here and now in the given conditions of the present. (1958/1949: 16)

Buber suggests that this theory and practice developed through the work of three generational pairings of activist thinkers: Saint-Simon and Fourier, Owen and Proudhon, and Kropotkin and Landauer. According to Buber, Saint-Simon contributed the insight that it was neither necessary nor desirable to organize modern industrial societies along the lines of a dualistic split between the social and the

91

political, between those who produce and those who lead. Instead, he proposed a unitary social order in which those who served certain functions led in the organization of those functions. Fourier and Owen contributed the observation that this was possible only when production and consumption are intimately linked, as they are in small, self-sustaining communities. But, Buber argues, in the work of Fourier and Owen, these communities remain disparate; it was Proudhon's contribution to show how they might be combined into non-statist federative structures. Kropotkin further developed Proudhonian federalism by showing how to promote and organize it, and by arguing for the presence of a universal principle that would drive it, that is, the principle of mutual aid. Finally, Landauer is credited with the realization that no revolution is necessary to begin constructing the new world in the shell of the old—if mutual aid is always with us a principle, then socialism can be created, for those who choose it, at the time and place of their choosing.

The second thread holding together this discussion is Marx and Engels' critique of Utopian socialism which, along with the takeover of the First International, marks a crucial moment in the establishment of the hegemony of hegemony within socialist theory and practice. Since its invention in the mid-nineteenth century, the category of Utopian socialism has allowed marxists and liberals alike to dismiss anarchism without any real engagement with its theories and accomplishments. Thus, while the rancour between Marx and Engels and those whom they called Utopian socialists is well known, it is important to return to the texts which preserve their polemics so as to challenge received readings that may hide more than they reveal. For example, it is rarely noted that Marx and Engels were not entirely dismissive of their Utopian contemporaries. Engels had praise for Saint-Simon's contributions to the nascent science of political economy (Engels 1978/1880: 689) and lauded Fourier for using the dialectical method 'in the same masterly way as his contemporary Hegel, while managing to jettison the teleological and eschatological elements of Hegel's historical narrative'. To Robert Owen, Engels attributed an early appreciation of the importance of both the labour theory of value and the concept of surplus value (690–2). Marx, for his part, even after *The Poverty of Philosophy* had appeared, continued to be of the opinion that Proudhon's first book, *What is Property?*, was 'epoch-making', and in a characteristically masculinist gesture, praised its 'strong, muscular style' (Marx 1975/1847: 179–80).

Despite their occasional kind comments, it is none the less clear that Marx and Engels believed that certain of their precursors had outlived their usefulness, had come to represent fetters, if you will, on the forces of socialist intellectual production. One problem they identified was the belief in total, instantaneous revolution. Utopian socialists did not 'claim to emancipate a particular class to begin with,' but wanted to liberate 'all humanity at once' (Engels 1978/1880: 685). The second problem with Utopian socialism, according to Marx and Engels, was its reliance upon rationalistic social experiments rather than situated analyses of historical, political, and economic conditions. As such, Marx declared, Utopian socialism 'tries to impose new hallucinations and illusions on the people instead of confining the scope of its knowledge to the study of the social movement of the people itself' (Marx 1978/1874–75: 546). The biggest problem of all, though, was that these two tendencies worked together to create an eclectic 'mish-mash' (Engels 1978/1880: 694) of theories, rather than a single, objectively verifiable socialist narrative.

What, then, are we to take from the marxist critique of Utopian socialism? To answer this question, it is necessary to look more closely at what the Utopian socialists in particular, and anarchists in general, actually have had to say with respect to the transition to a socialist society. There is much more there than Marx and Engels ever imagined, or were willing to admit, and much that has been added to the literature since their time. It must also be noted that the ignorance and dismissal are mutual—anarchists could also benefit from closer attention to certain streams of contemporary marxism, particularly its non-leninist autonomist variants. And finally, classical and contemporary anarchisms and marxisms can and should be overhauled, in fact *are being* overhauled, under the influence of recent trends in social, political and cultural theory. What is necessary, then, is a history of the present of the logic of affinity that not only works across the long-standing split between anarchists and marxists, but immanently critiques, from a twenty-first-century perspective, certain values and assumptions that they share.

In this discussion I will also be guided by the re-readings of classical anarchism that have recently been undertaken by writers such as Todd May (1994), Lewis Call (2002) and Saul Newman (2001). These forays into the field of postmodern/poststructuralist/or simply postanarchism (Adams n.d.) advance a critique of the rationalism and humanism of the early anti-authoritarian socialists without dismissing them entirely as relics of a bygone era. Rather, postanarchist theory claims that certain elements of the Nietzschean-inflected thought of Deleuze and Guattari, Foucault and, in some cases, Lyotard and Baudrillard, can be seen as distinct from, but compatible with, an anarchism stripped of its essentializing elements. Call, for example, claims that 'Foucault's postmodern anarchism', while it exhibits some commonalities with the classical theorists, 'is ... very different from the merely modern anarchism of Bakunin or Kropotkin' (2002: 65). While I would generally agree with this assessment, there are two problems with the general drift of postanarchist theory that I want to address. First, and as Call notes, Foucault never allied himself with any of the 'isms', refusing all such categories as instruments of semantic and political policing. So, calling him an anarchist *post-mortem*, even with an appropriate warning, is to do violence to his writing and activism; it is precisely to attempt to hegemonize Foucault in the name of non-hegemonic struggles. Or, as Newman says of his own project, it is to make 'shameless use of [post-structuralist] ideas to advance the [anarchist] argument' (2001: 7).

Similar problems arise in trying to appropriate Nietzsche, Deleuze or Derrida into the anarchist canon. But still, there is something to this desire to bring these writers on board, something that makes sense theoretically and politically. I would say it is this: that there exist certain common themes and ethico-political commitments between anarchism and poststructuralism, such that one might accurately point out certain *anarchistic elements* in the work of those late twentieth-century French writers who took up and deepened Nietzsche's critique of western humanism and the project of the Enlightenment. Seen in this way, the project of postanarchism makes a lot more sense to me.

The second problem I have with postanarchist texts is their tendency to give a little less credit to the classical writers than I would find justifiable. Here I am keying on the phrase 'merely modern' in the quote from Call cited above. Working immanently within a tradition—working deconstructively or genealogically, for example—means paying attention not only to breaks and ruptures, but to the ways in which one form emerges out of another, the ways in which 'the new' never entirely displaces 'the old'. We can see this kind of logic in Derrida's reading of Lévi-Strauss, which inaugurated poststructuralism as such (Derrida 1978), or in any of Foucault's genealogical studies. In this book I will not have time to delve into the question of the extent to which Deleuze, Foucault or Derrida might owe certain unacknowledged debts to the anarchist tradition, debts incurred through a combination of an allusive style and a tendency (always resisted, of course) to reject anything that might look oh-so nineteenth century. It will have to suffice to show how the anarchist tradition has always been deconstructing itself, how classical anarchism was not simply, solely and unproblematically a rationalistic humanism with a vision of totalizing revolution leading to a transparent society. This tendency most certainly existed, and it was undoubtedly dominant. But along with it there were moments of awareness of the tensions and contradictions that are now being exploited by postanarchist re-readings. I cannot hope to show this with respect to all of the possible lines of critique of the classical writers. I will endeavour, however, to articulate how the logic of affinity has been always already present in anarchism, how it has existed as a counter-pole to the totalizing revolutionary urge that dominated not only anarchist socialism, but every other political ideology of the modern era as well. I will argue that, rather than singling out anarchism as particularly guilty of the sin of reproducing modernist humanism, we should see it as a particularly fruitful ground for displacing it.

This is, of course, ultimately what the most interesting postanarchist texts are trying to do, and if the readings produced by this confluence of traditions are a little hegemonizing and inadequately immanent, they are in good company. In closely reading the texts of the anarchist canon, one cannot help but notice that it is extremely common for anarchists simply to highlight those aspects of their predecessors' thought that please them and hide those which do not. It could be argued, of course, that this is what immanent critique does anyway. The difference would be—and it is a crucial difference—that the immanent critic tries to proceed with her eyes as open as possible, demanding that her readings make sense not only on the basis of a single text or passage, but in the ever-widening contexts into which the texts under study might be placed. The immanent critic—the genealogist—is also extremely sensitive to how concepts and the discursive systems in which they circulate shift over time. All of these considerations require that one proceed very carefully when developing concepts and linking writers into theoretical traditions so that, for example, one does not mistake an obvious technocratic liberal like Saint-Simon for a socialist on the basis of one fragment from his late writings. This method of productive misreading in the interest of producing a profound lineage is rampant within the anarchist canon, perpetuated by a tendency of later generations to accept the pronouncements of their predecessors without bothering to read the relevant texts for themselves. By revisiting the classics, by reading them carefully using the tools made available by marxist and poststructuralist critique, postanarchism opens up the possibility of revising and rejuvenating the anarchist canon, and perhaps even toppling one or two tottering statues along the way.

First things first though. I want to turn now to an in-depth discussion of Utopian socialism, judging the applicability of Marx and Engels' critique not only to the trio to which it was originally addressed, but to other key figures in the classical anarchist canon to whom it has subsequently been assumed to apply: Godwin, Proudhon, Bakunin, Kropotkin and Landauer. Along the way I will pay particular attention to how the logic of affinity has struggled to emerge from underneath the dead weight of the hegemony of hegemony.

WILLIAM GODWIN: THE RATIONALIST WHO WOULD MAKE NO PROMISES

Although Godwin was a contemporary of those whom Marx and Engels called 'the three great Utopians' (Engels 1978/1880: 685) he was much more successful in evading the sharp point of their polemical pen. In a letter to Marx in 1845, Engels declared that despite its 'many excellent passages', Godwin's *Enquiry Concerning Political Justice* was 'altogether distinctly anti-social in its conclusions', and thus did not merit close attention (1982/1845: 26). It would appear that Marx and Engels considered that Godwin was not enough of a socialist to constitute a threat to their doctrines. But his rejection of the state form and the social contract that binds us to it has earned him pride of place as 'the first theorizer of Socialism without government—that is to say, of Anarchism' (Kropotkin 1912: 13). Much has been said on this account, so I will not say more. Rather, I will focus on how Godwin's work is relevant to the struggle between the logics of hegemony and affinity within the modern socialist project as a whole, and to the emergence of the strategy of structural renewal within anarchism in particular.

As is well known, Godwin did not accept the coercive aspect of state rule on the grounds that the use of force 'puts a violent termination upon all political science; and seems intended to persuade men [*sic*], to sit down quietly under their present disadvantages, whatever they may be' (1993/1793: 140).[1] This termination of debate, of course, is precisely what is valued by classical liberals and the majority of marxists, and it is not surprising that Godwin, as a rationalist above all, refused to accept it as a founding principle of a just society. But it is less often noted that he was also critical of what would later be called the consensual aspect of hegemony, holding that 'acquiescence is frequently nothing more than a choice on the part of the individual of what he deems the least evil' (144). Lack of opposition, he argues, cannot and must not be construed as consent, especially where the individuals supposed to have consented are unaware that they have done so. But even when consent is actively and freely given, Godwin questions, in a decidedly Nietzschean fashion, whether a decision to abide by a certain set of rules at a particular point in one's life should bind one forever. Is it not possible that we may change, or the society may change, and if so should we not be able to withdraw our consent? For Godwin arrangements between individuals must be *finite and limited* if they are to be just. Since the state demands a commitment that is infinite and unlimited, he rejects its legitimacy and authority.

This is not to say, however, that Godwin saw human interaction as nothing more than a summation of particularistic interests. Rather, and despite Engels' protestations, the principle of justice that is consistently invoked as the final arbiter in all of his deliberations is a *collective* principle. For Godwin, 'justice is a general appellation for all moral duty', and his own enquiry is cast as a contribution to the 'science of morals' (80). This is to say that he saw himself as operating squarely within the tradition of rationalistic moral theory; he thought it should be possible to devise a framework that would enable one to discriminate clearly between choices which 'must be either right or wrong, just or unjust' (80). And how are we to tell the right from the wrong? Godwin declares that 'if justice have any meaning, it is just that I should contribute everything in my power to the benefit of the whole' (80). It is in its orientation to 'the whole' that Godwin's conception of justice takes on a hegemonic, universalizing tone that Marx and Engels might have found attractive. 'There is scarcely any modification of society but has in it some degree of moral tendency', he notes. 'So far as it produces neither mischief nor benefit, it is good for nothing. So far as it tends to the improvement of *the* community, it ought to be *universally* adopted' (89; italics added).

This formulation would certainly have been seductive to the Enlightened mind, but it begs the question of how this whole, 'the community', is to be conceived? What are its limits? Where does it begin and end? Who is part of it and who is not? And among those who are within, are there some who are more in than others? If he were indeed some kind of socialist, we would expect Godwin to be an advocate of equality of condition within 'the whole', to reject as inadequate the equality of opportunity that suffices for liberal theorists. But his position on this issue is neither liberal nor socialist—it is in fact aristocratic. While Godwin was of the opinion that 'we are partakers of a common nature' (105) and thus are moral and physical equals in our *potential* capacities, he also believed that 'the treatment to which men [*sic*] are entitled is to be measured by their merits and their virtues' (107), that is, according to their *actual* capacities, however these might be measured and judged. In a famous thought experiment designed to illustrate the universality and impartiality of his conception of justice, Godwin suggests that if the palace of the Archbishop of Cambray were in flames, and one were faced with saving either him or his chambermaid, 'there are few of us that would hesitate to pronounce ... which of the two ought to be preferred' (81). Most who call themselves anarchists would choose the chambermaid, but Godwin comes down on the side of the church bureaucrat, claiming that he is more likely to do more good for more people, and that 'being possessed of higher faculties, he is capable of a more refined and genuine happiness' (81). It is probably this aspect of his thought that caused Engels to reject Godwin as a potential interlocutor.

At the same time, however, his vision of a future society is clearly relevant to the genealogy of the logic of affinity. He was an opponent of representative national government, holding that this form created a 'fictitious unanimity' (568) that amounted to a tyranny of the majority, or worse, a *real* unanimity based on self-interest and rhetoric rather than reasoned argumentation. (Those of us who live in mass-mediated societies are all too familiar with the latter case.) His vision of an alternative society was decentralized and based on small, face-to-face groupings (the parishes) which would form a loose 'confederacy' (576). Unlike nation-states, he argued that freely federated parishes would have no interest in extending their territory by force: 'If we would produce attachment in our associates, we can adopt no surer method than that of practicing the dictates of equity and moderation' (566). Most issues arising between the federated districts would be settled informally, or by arrangements that 'need not be in the strictest sense of perpetual operation' (566).

Having reduced the national assembly to something that convenes at the will of the parishes under conditions of strict necessity and is dissolved until needed again, Godwin goes on to consider the kind of authority such an assembly should have. 'Are they to issue their commands to the different members of the confederacy? Or is it sufficient that they should invite them to co-operate for the common advantage, and by arguments and addresses convince them of the reasonableness of the measures they propose?' (576). One might expect him to come down unequivocally on the side of reasonableness, but once again he defies expectations. 'The former of these would at first be necessary', Godwin argues. 'The latter would afterwards become sufficient' (576). Here Godwin encountered a difficulty that was to recur in anarchist theory and practice: we can call it the problem of the Revolution, the apparent need for a radical break, a discontinuity at the same time temporal and institutional-subjective, between the 'bad' social totality of today and the 'good' one that lies just around the corner. Like so many who came after him, Godwin assumed that people living under state-capitalist institutions were too corrupted by those institutions to be able to make a new society immediately. They would not be able to see reason, and thus would have to be forced into a state where they could. It is curious, but characteristic of many classical socialists, that he was not troubled by the end of political debate that would be brought about by the use of force in *this* context—it is, after all, force used in the pursuit of *what he sees as* reason and the good life.

Godwin's handling of the problem posed by 'insufficiently developed' human beings is typical of both his time and his temperament. It did not occur to him that his universalizing doctrine could be responsible for producing the very lack it found in others. Rather, the lack is seen as inherent to the subjects themselves, and thus thought to be remediable by action upon them—in this case, via education. Thus Godwin believed that, with sufficient instruction, chambermaids and others of inferior quality might be 'roused from the slumber of savage ignorance' (180), brought to see the light of reason and rendered able to understand the single and uniform truth to which God(win) was leading them. Human nature may not be perfect, but it *is* perfectible, though in human affairs 'everything must be gradual' (182). Thus we might look forward to the eventual 'dissolution' of political government, that is, to the withering away of even the periodic national assemblies and inter-parish juries. Once the appropriate institutions were in place, Godwin maintained, 'the whole species will become reasonable and virtuous' (577).

Godwin's perfectionism has been much remarked upon in its historical context, but its influence upon contemporary anarchist theory and practice has not been so widely discussed. Is it possible, perhaps, that the difficulties later anarchists have had in appealing to workers, people of colour and women are related to a lingering subterranean flow of Eurocentric masculinist rationalism and perfectionism? That the founding thinkers of anarchism have often suffered from some combination of misunderstanding, ridicule, imprisonment and assassination cannot be denied. But is it possible that in attempting to produce friendly readings of these marginalized figures, some problematic aspects of the anarchist tradition have not been adequately understood or criticized? Certainly it seems clear that Godwin's thought was rife with contradictions: a humanitarian and a libertarian, a hegemonic thinker who recognized the promise of the logic of affinity and free federation, an individualist and a moralist, an unrepentant rationalist who refused to make promises. Approached in a spirit of immanent critique, his work has enduring value precisely because it contained many of the conflicts that are still being played out within the anarchist tradition, sometimes through direct lines of descent, and sometimes through more tenuous patterns of observable regularity.

THE UTOPIAN SOCIALISTS PROPER: OWEN, FOURIER AND SAINT-SIMON

Both kinds of links seem to have existed between Godwin and the socialists Marx and Engels loved to hate, although it would seem that only one—Robert Owen—was directly influenced by his work (Marshall 1992: 390). Certainly they were all, to some extent, contributors to a diverse and developing set of ideas about social change. Anna Wheeler and William Thompson, the authors of the *Appeal of One Half of the Human Race, Women, against the Pretensions of the Other Half, Men* (1970/1825) were not only familiar with *Political Justice*, but Wheeler had met Fourier and a group of Saint-Simonians in France, and translated Owen's French correspondence. There are also many points in their texts where they comment upon each other's work, either directly or through allusion. More important than these interpersonal connections, however, are the theoretical and practical tendencies that unite and divide these figures who have, I would suggest, been forcibly placed in the same category by both their marxist detractors and anarchist defenders.

Without a doubt, Owen and Saint Simon were, like Godwin, Utopian thinkers in the loose sense that they were rationalist-perfectionists who believed in the coming of a society in which domination and exploitation would be entirely eradicated. Owen believed he had discovered the key to nothing less than 'the permanent rational system of society, based on the ascertained laws of nature', which would 'root up and utterly destroy the old vicious and miserable system', bringing a change so profound that 'competition, strife, and wars, will cease forever' (Owen 1973/1849: 56). Although he was famous for his experiments, they were intended only as trials of what he already knew to be the one correct way in which the good society was to be ordered. Once this was attained, there would be no further need for pursuit of alternatives. Saint-Simon had a similar vision. He thought that 'permanent public peace' and 'individual and collective happiness' would be achieved once 'the most important industrialists are in charge of the administration of public wealth' (Saint-Simon 1976/1823–26: 183). Fourier also tended in this direction; he was, as Hakim Bey has pointed out, a 'logothete' (Bey 1991c) who looked forward to 'an unbounded philanthropy, a universal good will' (Fourier 1971a: 61). At the same time, however—and unlike Owen and Saint-Simon—he cautioned that 'we must not persuade ourselves that in Harmony mankind are brothers and friends. It would be robbing life of its salt to cause the shades of opinion, contradictions, antipathies even, to disappear from it' (1971a: 159 n. 1). Granted, Fourier valued discord only between what he called 'series', and not between individuals. But by giving what he called 'cabalism' a secure place in the world of Harmony he opened up a path that is crucial to the emergence of the logic of affinity, and must be noted as a deviation from the common caricature of the Utopian socialists as advocates of an entirely transparent society.

Engels himself, of course, at times professed a belief in the eventual withering away of the state and the coming of a classless society, and was, like Marx, Utopian in this sense. His problem, as I have noted above, was not so much with the assumption that a new era was coming, but with the way in which his predecessors hoped to hasten its arrival: they did not seek to emancipate a particular class to begin with but wanted to liberate all of humanity at once. This is a complex claim which needs to be examined in terms of its three components: the class analysis, or lack thereof, which drives the thought of Saint-Simon, Fourier and Owen; related to this, their desire to liberate all of humanity; and, finally, their desire to do so 'at once'.

Engels' belief that a particular class needed to be emancipated first is based on the assumption that class inequality and therefore class struggle are unavoidable in historical (non-transparent) societies.

He is in fact accusing the Utopian socialists of not being adequately revolutionary, a charge which is difficult to refute since Saint-Simon, like Godwin, clearly wished to be seen as a reformist. 'Far from advocating insurrection and revolt,' he proclaimed, 'we are putting forward the only way to prevent the acts of violence which threaten society' (Saint-Simon 1976/1823–26: 185). He believed that 'the leading industrialists' should take control of the administration of public wealth, but only by peaceful means. Indeed, although he bequeathed to marxists and anarchists alike the notion of government as the administration of things rather than of people, and is often cited as one of the first proponents of a stateless society, his vision was hardly socialist at all. It was if anything libertarian, or what we would today have to see as *neo*liberal, in character. 'What the nation wants principally is to be governed as cheaply as possible' (184) he declared again and again, looking forward to a day when the function of government would be restricted to 'preventing the disruption of useful work' (105). Would Pinkerton cops and Plan Colombia be going too far in this quest, or just far enough? Saint-Simon also argued that during the French Revolution it was 'the industrialists' who suffered the most, since they 'twice lost their capital' (127)! From any number of textual indications of this sort, it seems clear that his conception of 'the industrial class' did not adequately differentiate between bourgeoisie and proletariat, and that those whom he expected to lead the coming reform were the owners, not the workers.[2] Although they were more critical of capitalism and its agents, Owen and Fourier also believed that the new age could only be ushered in by peaceful means. Owen insisted that revolutionaries were 'irrational' (1973/1849: xxiii) and Fourier assured his readers that the coming of Harmony would 'in nowise disturb the established order' (1971a: 66). Therefore, it has to be acknowledged that Engels was correct in asserting that these writers were not interested in emancipating a particular class by violent means.

What about the charge that the Utopian socialists foolishly set out to address all of humanity at once? Engels' tone here is, of course, characteristically mocking, and he is referring primarily to the lack of what he would have seen as an appropriately dichotomous class analysis, coupled with an ignorance of specific conditions in different countries. But his remark can also be taken quite literally. Fourier believed that once people saw how well his system worked, it would 'spread suddenly and spontaneously over the whole of the human race' (1971a: 50). For Harmony, as the only mode of social organization in alignment with the will of God, 'must extend and be applicable to all nations ... every people, age and sex' (48). Owen had a similar vision of a new world order which would 'extend over Europe, and afterwards to all other parts of the world, uniting all in one great republic, with one interest' (1973/1849: 65). On this point Saint-Simon was also vulnerable. He was wont to compare his native land with England, whose constitution he believed was founded on 'what is universally valid for all times and places, on what should be the basis of every constitution, the freedom and well-being of the people' (1976/1823–26: 84). He was an advocate of a common European constitution (85), and argued forcefully that his New Christianity should aspire to the status of 'the general, universal, and sole religion ... organizing the whole human race' (1975/1825: 298). Again, it would seem that Engels' assessment was to a great extent justified.

However, none of the three writers we are discussing argued that this peaceful global transformation would occur in one fell swoop. Saint-Simon, as I've already pointed out, was a reformist who didn't think any kind of radical transformation could, or should, happen quickly. Fourier and Owen also believed in peaceful means of social change. However, they saw the necessary transformation coming about in a way that was unrecognizable to both Saint-Simon *and* Marx and Engels. In contrast to the hegemonic visions of reform and revolution, Fourier and Owen assumed that voluntary associations would slowly but surely replace existing structures. Like Godwin, Owen based his system on townships full of properly educated subjects, which, as they increased in number, would choose to be 'federally united ... formed in circles of tens, hundreds, and thousands, etc.' (1973/1849: 65). Fourier saw the new world order arising out of imitation rather than education, as more and more people recognized the superiority of the social institutions of the Harmonic era and took it upon themselves to construct their own phalansteries. He also imagined a federative system, with the phalanxes organized into unions, districts, provinces, nations, and so on, within the 'spherical unity of the human race' (Doherty 1968/1851: xxxi). It is unclear, however, whether there would be any choice in this matter. Could a phalanx opt out of its union, a union out of its district, in the way that individuals could opt out of series within individual phalanxes? The existence of a 'world-wide corps of paladins, officers of the emperor and empress of unity' (Fourier 1996/1808) suggests that secession might be met with forceful efforts to maintain unity.

On this point, Buber holds that Fourier did not, and in fact could not, theorize harmony as a state pertaining between the basic units of association themselves:

> Each unit is a world on its own and always the same world; but of the attraction which rules the universe we hear nothing as between those units, they do not fuse together into associations, into higher units, indeed they cannot do so because they are not, like individuals, diversified, they do not complement one another and cannot therefore form a harmony (Buber 1958/1949: 20)

It would appear that Buber was unfamiliar with Fourier's writings on the ascending units of global unity, but his critique stands none the less.[3] In a desperate attempt to avoid the hegemonic moment through the rational ordering of all possible difference, Fourier fell prey to the logic of integration, that poor cousin of affinity and best friend of hegemony. A similar point could be made with regard to Owen, in whose work the logic of affinity also struggled with the hegemonic imperative, which appears in all its glory as the federation of townships approaches the limit point of the global republic, leaving no room for any other ways of life, all of which must by necessity be inferior and irrational to a life guided by the principles Owen had 'discovered'.

In pointing out these difficulties—and in accepting certain aspects of Engels' critique of Utopian socialism—I do not want to be read as suggesting that Saint-Simon, Fourier and Owen should be discarded, that they made no advances or have no relevance today. Rather, my goal is to show, first, how both the logic of hegemony *and* the logic of affinity are at work in their texts, and second, how an inability to adequately understand how these logics *interact* colours the dominant readings of the Utopian socialists. A similar re-reading also needs to be undertaken with regard to the ways in which these three writers handled the questions of democracy, equality and solidarity. For, on these points as well, it seems that certain crucial differences have been ignored and some common traits effaced in order to construct a unified field of Utopian socialism. Saint-Simon, as I have pointed out, did not reject government outright; he only sought to minimize its functions. His vision was also strictly anti-democratic and hegemonic in that he believed that 'the industrial class should be established as the first of all the classes', and 'the other classes should be subordinate to it' (1976/1823–26: 186). Owen was much more critical of capitalism than Saint-Simon, of course, and knew from his own experience the problems that attended the expert management of industrial relations. 'The people were slaves at my mercy', he noted after 30 years of his New Lanark experiment, 'liable at any time to be dismissed; and knowing that, in that case, they must go into misery, compared with such happiness as they now enjoyed' (1973/1849: 21). But he too did not reject government or the state form as such. His paradise on earth in fact assumed a clear distinction between government and governed; in it, the state would 'devise and execute' what it saw as being necessary for 'human happiness' (58). He also saw no need for democracy, participatory or even representative. Rather, he imagined that decisions could be made for everyone by those who were between the ages of 30 and 40, for local matters, or between 40 and 60 for issues external to the township. These ageist elites would have 'full power of government in all things under their respective directions, so long as they shall act in unison with the laws of human nature' (67).

Owen's lack of faith in youth is just one aspect of a general tendency to devalue the popular and the uneducated. When he first set out to transform the mill at New Lanark, he found many 'disadvantages in the character of the population ... their habits were intemperate, immoral, dirty, and most inferior' (1973/1849: 11). However, one way in which Owen represents an advance on Godwin and Saint-Simon—and in which he proves himself to be some kind of socialist, if not an anarchist—is his belief that every member of every township should possess 'similar advantages', with the only divisions being those of age (as already seen in his conception of government by the eldest). Thus, those at the same stage of life would have similar accommodations and access to goods and services. 'Perfect equality throughout life is the only foundation for a certain bond of union among men [sic]', he argued, 'and for an elevated state of society' (1973/1849: 123). Fourier, of course, begged to differ. He held that 'the associative regime is as incompatible with equality of fortune as with uniformity of character', and derided 'Owen's scheme of communism' as 'a mask for party spirit, a veil to cover the secret plan which tends to destroy the clergy and religion' (1971a: 128). Fourier also considered the holding of common property to be an idea 'so pitiful that it is not worthy of refutation' (128). Within the phalansteries he argued that there must be differences of fortune ranging from abject poverty to the possession of 'hundreds of millions' (125), and held that those with more money were a 'higher' type than those with less. Political equality was also anathema to Fourier—in the local communities there would be 'heads', or monarchs, and each of the twelve levels of federation had an associated duarch, triarch, and so on, up to the omniarch who would govern the entire planet (Godwin 1972/1844: 76). Like Saint-Simon, he was no anarchist socialist in any sense we would recognize today, but what might best be called an authoritarian communitarian capitalist.

Despite these propensities, there are several aspects of Fourier's thought that merit attention. He was the only one of the so-called Utopian socialists, for example, to take up a radical critique of the subjection of women, devoting a section of the *Four Movements* to the 'Degradation of Women in Civilization':

> Is there a shadow of justice to be seen in the fate which has befallen them [women]? Is a young woman not a piece of merchandise offered for sale to whoever wants to negotiate her acquisition and exclusive ownership (1996/1808: 129)

Fourier goes on to link progress in the emancipation of women with progress in general towards the higher forms to come after Civilization has been superseded, declaring that 'all the periods which produce social well-being ... should have no pivot ... except the progressive liberation of the weaker sex' (92). Here, as elsewhere, Fourier shows signs of sexist tendencies at the same time as he is critical of the position of women in Civilization. But he did make a *beginning* that was uncommon for his time and place. He was similarly advanced in his opinions on homosexuality, according all of what he called 'the natural manias' an equal place in 'the omnigamous series' (Fourier 1971b: 348). Interestingly, this was not based on mere tolerance. 'Since love tends to manifest in manias,' Fourier argued, 'and since all manias will be highly useful in Harmony, their development will be systematically encouraged' (349). Anticipating Freud's theory of repression, Fourier was concerned that if someone who was 'born to be a hair-plucker or a heel-scratcher in love' was not able to satisfy his desire due to social stigma, 'he will succumb to other, harmful manias' (354). In order to encourage others to reveal and revel in their own preferred forms of sexual activity, Fourier declared that he himself was prone to a 'mixed extra-mania: sapphianism or the fondness for lesbians' (349).

Fourier's work is also crucial to the development of the logic of affinity. Although he was clearly a rationalist, Fourier was decidedly *not* a moralist—he held morality to be 'the mortal enemy of passionate attraction' (1971a: 55). Within particular communities he argued that all relationships should be based on the series, which are formed according to 'affinity of tastes' (122). Each group must be composed only of such members as take part 'passionally, without having recourse to the mediums of necessity, morality, reason, duty, and compulsion' (159). Unlike all of the writers I have discussed so far, he valued spontaneity and passion over abstract conceptions of duty, and argued that in the age of Harmony, what he called 'traction', or natural authority, would replace abstract conceptions of duty and hierarchical constraints—again, though, perhaps only within the phalanxes and not between them. Despite his critique of moral laws, however, he was unable to free himself from a reliance upon a dubious hope to unite Christianity and natural science. Like Saint-Simon and Owen, he tried to show that his theory was based on a universal natural/divine law: God, he argued, 'rules the universe *by Attraction and not by Force*' (61; italics in original). He was endlessly critical of 'the sophists', 'the philosophers', because their doctrines were not in line with the laws of God and Nature. He also felt, rather ironically given the place he has been allotted in the history of socialism, that his discovery of the law of passionate attraction put him above those 'utopia-makers' (56) who fail to properly observe how the natural, human and divine worlds participate as one entity in a glorious dance of passionate attraction. Fourier's struggle to justify his commitment to spontaneity and situated ethics provides another example of the complex ways in which the logic of affinity is enmeshed with the logic of hegemony in his texts.

This complexity, I would argue, has been unfortunately obscured by the readings of Marx and Engels, and by later commentators following their lead, who have tended to view 'the Utopian Socialists' through the lens of a single dismissive label. The dominant reading has also tended to obscure, through a subtle redirection of the critical gaze, what can only be seen as Utopian elements in Marx and Engels' own formulations. From the vantage point of the end of the twentieth century, we can now see that Lenin's Soviet Union and Mao's China were in fact experiments, not objective demonstrations of the validity of the materialist conception of history. Thus it might be argued that actually existing socialism stands in the same relation to historical materialism as Brook Farm does to the theory of passionate attraction. Both were rationalistic experiments, differing only in the degree of their longevity and (very importantly) the amount of damage they did. Not only Saint-Simon, Fourier and Owen, then, but *Marx and Engels as well*, were under the spell of what Deleuze and Guattari call Royal Science—they tended to approach reality in terms of general theorems and axioms, rather than particular, situated problems. They believed that they had uncovered a fundamental law of human development that would make it possible to instantiate an ideal society, and wanted to instantiate that society as soon as possible. The differences between them appear only at the level of the content of the ideal and the way in which they thought it could be realized. These differences might seem small, but they were enough to ensure the predominance of a totalizing, class-based politics for the next 150 years. The task of developing a more coherent vision of non-hegemonic forms of social organization and transformation was not, however, abandoned. Anarchist theorists such as Proudhon, Bakunin, Kropotkin, Malatesta and Landauer continued to develop and critique the formulations of the Utopian socialists, pushing the theory and practice of affinity-based politics to greater levels of complexity.

ANARCHIST THEORY AFTER UTOPIAN SOCIALISM: PROUDHON

At this point I want to recall Martin Buber's argument regarding Proudhon's relation to Fourier and Owen. In the work of the latter two thinkers, Buber suggests, the local communities remain disparate; it was Proudhon's contribution to show how they might be combined into non-statist federative structures. The preceding discussion has shown that Fourier and Owen did in fact theorize the growth of larger structures out of the basic units of association. Does this then mean that Proudhon's work is less relevant for the genealogy of structural renewal than Buber claims? On this particular point I would say it does; and it must be noted that Proudhon also perpetuated other problematic aspects of Utopian socialism. He believed in 'the indefinite perfectibility of the individual and of the race' (1923/1851: 243), and declared that the 'historical law'(1971/1863: 34) he had 'discovered'—the principle of federation—was 'the one correct system' and represented 'the greatest triumph of human reason' (1971/1863: 5). Thus he uncritically reproduced the scientism, rationalism and perfectionism of his predecessors. Proudhon's texts do contain, however, some signs of struggle with the Sirens of the transparent society. On the one hand, he proudly states that under federalism 'all those partisan divisions which we imagine to be so profound, all those conflicts of opinion which seem insoluble to us, all those random hostilities for which there appears to be no remedy, will instantly find a definitive solution in the theory of federal government' (1971/1863: 7). But elsewhere he rejects this assumption, declaring with equal force that 'there has never been an example of a perfect community, and it is unlikely, whatever degree of civilization [is attained] ... that all trace of government and authority will disappear' (1969: 105). Proudhon was thus one of the first thinkers of the anarchist tradition to achieve, however fleetingly, the insight that it might not be possible to entirely eliminate domination from human relationships.

In his discussions of federalism, however, he rarely, if ever, talks about the imperfections that we can expect to linger. He focuses rather on the positive aspects of the New Dawn:

> The federal system is applicable to all nations and all ages, for humanity is progressive in each of its generations and peoples; the policy of federation ... consists in ruling every people, at any given moment, by decreasing the sway of authority and central power to the point permitted by the level of consciousness and morality. (1971/1863: 49)

This appeal to a universal progression—graded, of course, by the 'level of development' of a particular people—makes it clear that Proudhon never fully freed himself from the burden of the Hegelian conception of history that Marx and Engels were so happy to find in his work. But, even though he thought that 'historic evolution' was 'leading Humanity inevitably to a new system' (1923/1851: 126), he also appealed at times to 'the fecundity of the unexpected' (1969: 104), that is, to what poststructuralist theorists would later refer to as contingency or the event. Again, this is an early appearance of an element of anarchist thought that is crucial to the development of the logic of affinity. Like the questioning of the transparent society, however, and clearly related to it conceptually, the appeal to the unexpected surfaces in Proudhon's texts only to be rapidly submerged under the teleological flow of history towards the Absolute.

It is also important to address the question of how Proudhon thought that the sway of authority might be decreased. His thinking on this question also seems to have been contradictory. In *The General Idea of the Revolution in the Nineteenth Century*, published in 1851, he says that the 'the people' will have to start the revolution, by getting together and deciding to tell their representatives: 'we desire a peaceful revolution, but we want it to be prompt, decisive, complete'

(1923/1851: 174). 'You should make use of the very institutions which we charge you to abolish, and the principles of law which you will have to complete, in such a way that the new society may appear as the spontaneous, natural, and necessary development of the old, and that the Revolution, while abrogating the old order, should nevertheless be derived from it' (174). For the early Proudhon, then, the revolution was not a singular break, but a *process*, and he held to this formulation throughout his career. However, his conception of the role of 'the people' seemed to change. By the time of the *Principle of Federation* (1863), he seems to have lost his faith in both them and their representatives. The masses 'create absolutism' (1971/1863: 26) and demand authority, he declares. 'Left to themselves, or led by their tribunes', they will 'never create anything' (28). Indeed, he is of the opinion that one of the values of the federal system, as he proposes it, is that it 'puts a stop to mass agitation ... it is the end of rule by the public square' (62); it is 'the salvation of the people, for by dividing them it saves them at once from the tyranny of their leaders and from their own folly' (62).

This formulation begs the question of precisely how decisions will be made, if not by the people or their representatives. Curiously, for the first avowed anarchist, this function is attributed to the state. In Proudhon's federal system, it appears as 'the initiator and ultimate director of change' (45), the 'prime mover and general director' (48). To be fair to him, it must be noted that Proudhon was a strong partisan of the division of powers, and held that 'the role of the state or government is essentially that of legislating, instituting, creating, beginning, establishing; as little as possible should it be executive' (45). However, he does argue that after political change has been achieved, 'the federal government must necessarily proceed to a series of reforms in the economic realm' (70). Here, again, the state appears as the locus of social action. The centrality of the state form as a planning, directing and regulative power is an extremely problematic, but almost entirely overlooked aspect of Proudhon's theory of federalism. Especially when considered alongside his commitment to the division of powers, it shows that there is a strong liberal inflection to his thought, an inflection that we ignore only at our peril.

Of course, the state is not the sole actor in Proudhonian federalism. The federal bureaucracy is intended to act in concert with agro-industrial federations based on mutualism of credit. 'The purpose of such specific federal arrangements', he argues, 'is to protect the citizens of the federated states from capitalist and financial exploitation, both within them and from the outside' (1971/1863: 70). Just as Proudhon sought to minimize state domination to the level he thought was possible and acceptable, he also sought to ward off capitalist exploitation. Again, however, one might take issue with how far he was willing to go in this regard. As is well known, he envisaged three different modes of economic organization. For the peasants, freeholding in land, including the right of alienation of property; large-scale, complex enterprises would be worker-controlled, but would 'submit [themselves] to the law of competition' (1923/1851: 222) in their mutual relations, and though ownership was to be through equal shares, 'pay is to be proportional to the nature of the position, the importance of the talents, and the extent of responsibility' (222). As with his relation to the state form, we can see that Proudhon only sought to ameliorate some of the excesses of capitalist individualism and profit-seeking, not to do away with them entirely. Indeed, in the third case, that of small shopkeepers and manufacturers, he saw no need for association at all, even where individual proprietors took on journeymen and became employers (217).

Although he is much celebrated for his advocacy of decentralized forms of social organization, a careful reading of Proudhon's texts gives reasons for concern on this account as well. In the *General Idea*, written when he was very much under the sway of Saint-Simonian doctrine, Proudhon suggests that Saint-Simon's followers have interpreted their master incorrectly: Saint-Simon was always an anarchist, and if he used the term government, we are to take this 'as an analogy' only (1923/1851: 123). The core of Saint-Simon's argument, according to Proudhon, is that 'industrial organization' will not exist alongside of government, but will *replace* it (122). Replacing the state apparatus with an economic one, however, does not necessarily solve the problem of rational-bureaucratic domination. Indeed, Proudhon argues that 'in the place of political centralization, we will put economic centralization' (246). This aspect of his thought is once again in tension with an affinitive element: like Godwin, he consistently maintains that all relationships must be based on contracts rather than laws (1923/1851: 205–6; 1971/1863: 38). But here too there are problems, in that a focus on the contract and a desire to minimize state interference are quite in keeping with a libertarian capitalist position, that is, with the imposition of 'that economic unity which is destined to replace political unity' (247).

In fact, Proudhon's definition of anarchism, as it appeared two years before his death, in no way precludes this kind of reading. For him, in an anarchist society:

> political functions have been reduced to industrial functions, and social order arises from nothing but transactions and exchanges. Each may then say that he is the absolute ruler of himself (1971/1863: 11)

A certain sort of ethical aporia appears in Proudhon's contractual conception of how relations between individuals and groups are to be governed, a forgetting of the solidarity that is paramount to so many others who are included in the anarchist canon, and of the critique of the social contract inaugurated by Godwin.

Altogether, it seems to me that Proudhon cuts a rather ambivalent figure. While remaining a rationalist perfectionist with a teleological conception of history, he at least begins to see the impossibility of a world without relations of power and is aware of the fecundity of the event. At the theoretical level he doesn't add much that's new to anarchist discussions of federalism, and may even be said to have taken backwards steps in his statist, economistic, capitalistic and centralized vision. Thus there is some validity in the claim that he was, above all, a 'pragmatist' (Gambone 1996). Like Saint-Simon, Fourier and Owen, and despite his much-celebrated and truly wonderful rants against capitalism, the state form and modern life in general, Proudhon seems to have been driven by a desire make the best of existing institutions, rather than to replace them with alternative modes of organization. This reformist tendency is perhaps what has made it possible for his work to be appropriated by both socialist and libertarian anarchists, as well as by liberal theorists of federalism (LaSelva and Vernon 1998). But it also takes us away from the logic of structural renewal, which requires a different conception of the passage from present to future modes of existence, one that is neither reformist nor revolutionary. To arrive at this point of complementarity, however, it will be necessary to pass through both poles of the dichotomy. If Proudhon can be seen as representative of the reformist current in anarchism, then Bakunin, whom I will discuss next, might serve as an exemplar of its revolutionary counterpoint.

BAKUNIN AND THE SOCIAL REVOLUTION

Bakunin's work, like Proudhon's, was very much influenced by his predecessors. Echoing the perfectionism of Godwin, Bakunin declared that anarchists are 'enemies of all power' (1990/1873: 136) whose goal is the achievement of 'the most rational possible organization of social life' (133). Like Saint-Simon, Fourier and Owen, his vision was millennial and apocalyptic: he saw the Paris Commune as 'inaugurating [a] new era, that of the final and complete emancipation of the masses of the people' (1973/1871b: 199). At the same time as he reproduced much of what came before him, however, Bakunin made certain advances that are relevant to the emergence of the logic of affinity. The most important of these was his insistence on the distinction between social and political revolution. In his polemics with Marx and his followers, Bakunin associated what he called political revolution with the desire to wield state power as a weapon of the dispossessed. Social revolution, on the other hand, was about *breaking* rather than *taking* state power.[4] Bakunin was also highly suspicious of the prominent role attributed to individual actors by political revolutionary strategies. In a social revolution, he argued, 'the spontaneous action of the masses' should 'count for everything' and that of individuals for nothing (1973/1871b: 203). Political revolution, then, involves a mass being led by a small cadre of charismatic individuals who no sooner take power than they install a new order, a new state, a new domination. Social revolution, on the contrary, is a spontaneous uprising with no leaders or preformed goals, a passage to anarchism (the just society) through anarchy (disorder and chaos).

To understand why Bakunin insisted on this distinction it is necessary to engage in greater detail with his theory of radical social transformation. Although he postulated the Revolution as a millenarian break, he saw its transition occurring via a two-stage process: first, existing institutions of domination and exploitation had to be destroyed, so as to clear the way for a second period in which a new world would be built. He was adamant that these two stages had to be carried out in the correct order: 'The abolition of the Church and of the State must be the first and indispensable condition of the real emancipation of society; after which (*and only after which) it* can, and must, organize itself in a different fashion ...' (1973/1871b: 205–6; italics added).[5] As is well known, Bakunin was not in favour of partial measures of any kind. It would be necessary to overthrow '*all* the heavenly and earthly idols' in order to organize a new world on 'the ruin of *all* churches and *all* states' (197; italics added). However, after the stage of destruction had ended, he argued that reconstruction could be expected to go on for 'an indefinite period', until the dawning of the day when 'the triumph of [the principle of social revolution] throughout the world removes its *raison d'être'* (1973/1866: 64). So, although he is known primarily as a violent revolutionary, and was certainly dismissive of the expectation that radical change could be achieved by peaceful means,[6] his theory of social transformation in fact includes a period after the revolution in which reform becomes viable, leading finally to a period in which *neither revolution nor reform is necessary*. Thus Bakunin simultaneously deploys and confounds the revolution/reform dichotomy, thereby opening up—perhaps for the first time—the theoretical possibility of disposing with it altogether.

Despite his best efforts to avoid authoritarian practices, however, Bakunin's vision contained elements that were clearly hegemonic. A 'popular social revolution', he declared, 'destroys everything that opposes' its flow (1990/1873: 133). It is a totalizing global force, beginning with the free association of workers in unions and communes, which are then linked to span regions and nations, to culminate in 'a great federation, international and universal' (1973/1871b: 206). There are important differences, though, between Bakunin's world-transforming vision and those of Fourier, Owen or Saint-Simon. In *Federalism, Socialism, and Anti-Theologism* he complained that Fourier and Saint-Simon, despite their important contributions to socialist thought, were 'authoritarian' and 'prescriptive' (1973/1895: 100), 'doctrinaire' rather than 'revolutionary' socialists (99).[7] Bakunin then drops Proudhon onto the stage as the first truly anarchist socialist, whose doctrine was 'based on individual and collective liberty and upon the spontaneous action of free associations ... excluding all governmental regimentation and State protection' (100). This is not, as I have shown, what Proudhon actually advocated, and it may be that Bakunin should get much of the blame for installing this false but pure Proudhon in the anarchist canon.[8] None the less, Bakunin should be given credit for the notion that he stuffed into the head of his predecessor—the idea of a non-statist (but still hegemonic) mode of social transformation and organization.

Bakunin also contributed much to the anarchist conception of the relationship between social science and social change. Like every progressive thinker before him, and many after, he was committed to the basic tenets of Western reason and the Cartesian project. The 'mission' of science, he solemnly declared, is by observation of the general relations of passing and real facts, to establish the general laws inherent in the development of the phenomena of the physical and social world; it fixes, so to speak, the unchangeable landmarks of humanity's progressive march by indicating the general conditions which it is necessary to rigorously observe and always fatal to ignore or forget. (1973/1871a: 159)

What makes him different from those whom he called 'doctrinaire' socialists, though, is his belief that while 'science is the compass of life', it 'creates nothing; it establishes and recognizes only the creations of life' (160). That is, for Bakunin, everyday lived experience, passions, needs and aspirations must be our guide, rather than scientific abstractions. In fact, he argued that science is simply unable to proceed without 'flesh and blood': 'Abstractions advance only when borne forward by real men' [sic]' (162). So, while Bakunin hoped to be seen as a 'scientific' socialist, in the sense that he placed reasoned argumentation above theology, metaphysics and mere juridical right, he did not want to be one of those 'priests of science' to whom the people would become indentured just as soon as they freed themselves from Christian orthodoxy (1990/1873: 135).

But if science alone can tell us what it is fatal to ignore or forget, why would we not give to its proclamations—and its proclaimers— the highest attention and honour? Bakunin argues that one problem with scientists is that there are too few of them. Unlike Saint-Simon, he is extremely wary of expert knowledge held in the hands of a few: 'anyone who is invested with power by an invariable social law will inevitably become the oppressor and exploiter of society' (134). His deeper objection, though, comes from his faith in 'life', which he fears would 'dry up' if science were allowed to take the pre-eminent place, thereby turning human society into 'a dumb and servile herd' (135). Prophetic words indeed for those of us who have experienced the vicissitudes of scientific reason through the long twentieth century.

Bakunin's belief that 'life' must always be valued over 'thought' is also expressed in his faith in 'the people'. Indeed, it must be said that just as Saint-Simon placed the burden of social well-being on experts, or Fourier on a well-designed social structure, Bakunin had an extremely optimistic estimation of the spontaneous will of the masses. For him, we might say, what is Popular is Rational. His long-standing reluctance to propose a specific plan of revolutionary action, for example, is primarily based upon a belief that the ideal social organization cannot be deduced, but is immanent 'in the people themselves' (135). This is a profoundly democratic doctrine, but it begs the question of how we are to know who 'the people' really are and what they really want. Who is working in consonance with the 'real ideal', and who is imposing a mere dictate of abstract scientific reason? Bakunin's answer is that the will of the people will be expressed through the 'free unions of popular associations' that will comprise the 'bottom' levels of the global hierarchy in the new world order (136). To my knowledge he had little to say about how these associations were to be organized, how they would carry out their business, apparently leaving the ancient problem of 'how to decide how to decide' up to 'the people' as well.

This reticence is certainly understandable in someone who was trying to crawl out from under the dead weight of doctrinaire (Utopian) socialism. But it begs the question of Bakunin's conception of his own relation to that entity in which he placed his revolutionary faith. Did he see *himself* as one of the people, and if so, why did he categorically refuse to participate in any attempt to discover its will? It seems clear that while he was extremely careful to avoid 'speaking down' to the masses, Bakunin nevertheless maintained a position that was separate from 'life'. He saw himself, in fact, as existing *below* street level, working through secret societies to constitute an 'invisible collective power' that would 'guide' the coming revolution (1973/1870: 178). As much as he was repulsed by Marx and Engels' authoritarian-statist politics, Bakunin advocated his own brand of 'dictatorship', one 'without insignia, titles or official rights, and all the stronger for having none of the paraphernalia of power' (180). While he imagined that workers, after the revolution, would be able to spontaneously organize themselves into a decentralized, bottom-up anti-hierarchy of global proportions, he had little faith in the ability of 'discussions ... or popular assemblies' to bring on the blessed event. Rather, it would be necessary for 'a few allies, but good ones—energetic, discreet, loyal' (180) to 'arouse, unite, and organize spontaneous popular forces' (182) so that they did not emerge in a fragmented fashion, but as a united front able to overthrow the currently constituted powers. 'The people themselves', Bakunin argued in 'Revolutionary Organization and the Secret Society', 'because of their ignorance ... are unable to formulate and bind themselves to a system, and to unite in its name. That is why they need helpers' (188–9). These helpers must be committed, professional revolutionaries, willing to work without recognition, and to pay the ultimate price if necessary. They must operate among workers, peasants, itinerant thieves and other outcasts to create that much-needed link between 'the people's instincts and revolutionary thought' (190).

One cannot read Bakunin's description of the characteristics and tasks of the 'collective dictatorship' of secret societies without recalling the revolutionary actors who populate the stage of Lenin's *What is to be Done?* They too are professionals pressed with the task of organizing and educating the ignorant masses, except that they are working for political rather than social revolution, and their task is to quell spontaneous expressions of rebellion in the name of a supposedly higher plan, rather than to ride them for what they might be worth. These are, of course, important differences, but Bakunin's argument against the state form hinges on an identification of means and ends in revolutionary practice. An authoritarian revolution will reproduce authoritarian forms, Bakunin rightly points out; an orientation to heroic individual leaders will disempower the people by allowing them, once again, to avoid thinking for themselves. On this logic though, can we not expect that a revolution organized by secret societies would reproduce its *own* characteristics? Why would the secret societies, having done such a fine job of bringing about the revolution, decide to disband upon its arrival, at a time when a myriad of difficult and complex tasks would need to be faced? Wouldn't they offer precisely what was needed for the new phase of reconstruction, that is, continued leadership of the masses, who could not by any means be expected to have become scientists and committed professional revolutionaries overnight? Is it not necessary for Bakunin to be pitted against Bakunin, to denounce the fantasy of the withering away of the secret societies?

KROPOTKIN: EXPROPRIATION AND SOCIAL (R)EVOLUTION

Kropotkin might be seen as just the person to carry out this task, but this reading is complicated by the fact that he had much to do with the creation of the canon in which Bakunin's work finds so secure a place. His long-running article on anarchism in the *Encyclopedia Britannica* (1910–74) contains a genealogy that is very similar to the longer and more detailed discussion presented in *Modern Science and Anarchism*. In this text Kropotkin names Godwin as 'the first theorizer of Socialism without government', Fourier, Saint-Simon and Owen appear as 'the three founders of modern Socialism', and Proudhon is credited with laying anew the 'foundations of Anarchism', by coming up with a theory of federalism without knowledge of the work of Godwin (Kropotkin 1912: 13).[9] But it is Bakunin who establishes 'the leading principles' (62) of what Kropotkin calls 'modern anarchism', which emerges in 1864 with the founding of the International Working Men's Association. The most important of these principles is the primacy of social over political revolution, which doctrine Kropotkin appears to accept unmodified from his predecessor. And yet one cannot help but be struck by the fact that Kropotkin never talks about those fundamental elements of Bakuninist social revolution, the secret societies. He doesn't say a word either for or against them; they simply drop off the agenda as though they had never existed. While this rhetorical strategy is in keeping with the anarchist tradition's tendency towards tendential readings, it creates certain theoretical problems, such as: What takes the place of professional underground agitation in Kropotkin's vision of social revolution? How are the theory and practice of anarchist social revolution altered by this implicit shift in tactics? And finally, what are the repercussions of this shift for the logic of affinity?

The goal of social revolution, according to Kropotkin, and clearly following in the wake of Proudhon, is to create a society where 'the functions now belonging to Government would be substituted by free agreements growing out of the direct relations between free groups of producers and consumers' (1912: 64). Like Bakunin, he expected this change to come about via mass direct action in which 'the people lay hands upon property' and use it to meet their own needs without the mediation of state or corporate forms (1990/1892: 71). Kropotkin referred to this process as *expropriation*, and he had much to say about it throughout his career. In *The Conquest of Bread*, written in 1892, he sees expropriation as a singular event made possible only 'when the revolution shall have broken the power upholding the present system' (36), and the people have made 'a clean sweep of the Government' (60). Like Bakunin, Kropotkin seemed to assume that the old order had to be pushed aside before it would be possible to begin anything new. He also held that expropriation must 'apply to *everything* ... that enables any man ... to appropriate the product of others' toil' (53; italics added). He was very much against what he called 'half-measures' (54), since 'there are ... in a modern state established relations which it is practically impossible to modify if one attacks them only in detail' (54–5). Because of the resistance that could be expected, Kropotkin noted that expropriation would have to be violent—the peaceful means advocated by the Utopian socialists simply would not suffice (21). Thus expropriation/social revolution appears as a rapidly occurring, total transformation enabled by a spontaneous violent uprising and subsequent collapse of the currently existing order. It appears, that is, pretty much as it did with Bakunin.

In this same text, though, there are signs that this reading is not quite adequate. Kropotkin in fact approved, in many instances, of what would seem to be half-measures. For example, due to the different insurrectionary potentials of the towns versus the countryside, he expected major cities to go socialist quickly, but allowed that alongside the revolutionary urban communes there might exist towns and regions still 'living on the Individualist system' and remaining 'in an expectant attitude' towards the Revolution (1990/1892: 84–5). This apparent discrepancy can be explained by positing that when Kropotkin is arguing for a totalizing instantaneous transformation, he is talking about *particular* city districts, towns or rural regions, each of which he sees as rising on its own and fully 'sweeping away' the *local* manifestations of state and corporate rule. Within the broader context of the nation-state, however, he does not have the same expectations, and allows for the co-existence of capitalist and socialist forms. This is also the case at the international level, where he imagines that although revolution will 'break out everywhere', it will take on 'divers aspects; in one country State Socialism, in another Federation; everywhere more or less Socialism, not conforming to any particular rule' (1990/1892: 85). Half-measures, it would seem, are unacceptable *only at the local level*.

In later texts Kropotkin presents his position more clearly, and in a way that supports and extends this reading. Arguing explicitly against the marxist social democrats in 1912, he suggests that it is not enough to impose legal limitations upon capitalism. Rather, 'we must already *now*, *tend to* transfer all that is needed for production—the soil, the mines, the factories, the means of communication, and the means of existence, too—from the hands of the individual capitalist into those of the communities of producers and consumers' (1912: 67; italics added). The key words here are 'tend to' and 'now', for with them Kropotkin is arguing not only against the marxist reformers of his own day, but also against his own previous position. 'Tending to' transfer the means of production 'now' could mean nothing other than refusing to wait for 'the people' to rise spontaneously and fully sweep away existing forms; it necessarily implies an acceptance of half-measures *even at the local level*. Thus by the time of *Modern Science and Anarchism,* Kropotkin had fully reversed his earlier position. 'Each step towards economic freedom,' he now argued, 'each victory won over capitalism will be at the same time a step towards political liberty ... and each step made towards taking from the State any one of its powers and attributes will be helping the masses to win a victory over Capitalism' (1912: 81).

Once he had formulated the idea of proceeding immediately with partial expropriation, Kropotkin was logically compelled to allow for the fact that the revolution could no longer be thought of as an instantaneous transition at *any* level. Does this mean he became a reformer? No, for he still insisted that expropriation must take on a violent character: 'We know that this conquest is not possible by peaceful means. The middle class will not give up its power without a struggle' (1912: 88). And he was not by any means ready to give up on the idea of a totalizing instantaneous transformation occurring at *some point* in the future—he saw partial expropriation as part of a 'preparatory period', a 'period of incubation' (1912: 88), an evolution in preparation for the revolution. How long might this period last? Four years in some cases, 50 in others, one really couldn't say. But what is crucial is that Kropotkin succeeded, despite himself, in standing Bakunin on his feet, as it were, by suggesting that rather than waiting for the revolution to occur before beginning to build a socialist world, it was possible and desirable to create the relationships we desire immediately, in the world in which we find ourselves actually living.

The incitement to construct alternatives here and now raises the question of what is to be created and how we might go about creating it. Like Bakunin, Kropotkin was very much concerned to ground his theories in the scientific method. 'Anarchism is a conception of the Universe based on the mechanical interpretation of phenomena', he declared, 'which comprises the whole of Nature, including the life of human societies and their economic, political, and moral problems. Its method is that of natural sciences, and every conclusion it comes to must be verified by this method if it pretends to be scientific' (1912: 38). But he also shared Bakunin's belief that it was impossible—even for social scientists—to 'legislate for the future' (1912: 87). Once again, we see the characteristic anarchist attempt to elude the Scylla of Utopianism while not being engulfed by the Charybdis of Science. Like Bakunin, Kropotkin wanted to put all power in the hands of 'the people'; in the heady days of the First International, he recollects, '[w]e did not pretend to evolve an ideal commonwealth out of our theoretical views as to what a society ought to be, but we invited the workers to investigate the causes of the present evils, and in their discussions and congresses to consider the practical aspects of a better social organization than the one we live in' (1942/1899: 116). In his case, however, this desire was not frustrated by an inherent mistrust of what 'the people' might do—there is no reference to a programme of education in Kropotkin's texts, no recourse to experts, no dictatorship of secret societies. He sees himself as neither above nor below the crowd, but as *one of* 'the people'. His particular contributions happened to be scientific and theoretical, but that fact in no way lifted him above, or hid him below, the day-to-day struggle for social change.

As a self-professed social scientist, Kropotkin dealt with the general characteristics of a more desirable social order, rather than with the minutiae of daily life within it. This aspect of his approach is evident in *The Conquest of Bread*, where he has quite a lot to say about what he sees as the central principles of a successful post-revolutionary society. But it is also apparent in his commitment to the task of working out an anarchist ethics, which he took up in *Anarchist Morality* (1890), *Ethics* (1968/1924) and, most famously, in *Mutual Aid* (1989/1902). In this book Kropotkin sets out to show that his brand of anarchist science is not Utopian, in the sense of producing grand ideas out of thin air, but works—as Marx and Engels said it should—by 'building [its] previsions of the future upon those data which are supplied by the observation of life at the present time' (66). With this caveat in mind, he argues that the social principle that anarchists rely upon is not a mere fancy, but has always existed, and still exists in modern European societies, as a 'tendency to constitute freely, outside the state and the Churches, thousands upon thousands of free organizations for all sorts of needs' (66–7). Kropotkin's favourite examples were the Swiss cantons and other remnants of the village commune, as well as the networks of communication and transportation that spanned national boundaries and managed to operate quite successfully without centralized control. All of this must be attributed, he argued, to the 'creative, constructive power of the masses' (1912: 4), which has been responsible for inventing and maintaining new social institutions throughout human history.

But, just as the constituent power of mutual aid has always worked to create free relations of affinity, Kropotkin notes that there has always been an opposing tendency, one that works to appropriate new social forms and their products. Throughout Europe, he argues, most of what we know as the 'progress' achieved prior to modernity was achieved by disparate but similarly organized village communities. These entities were so advanced that 'the States, when they were called later into existence, simply took possession, in the interest of the minorities, of all the judicial, economical, and administrative functions which the village community already had exercised in the interest of all' (1912: 152). Kropotkin thus prefigures the notion of the state as an apparatus of capture, which would be worked out in greater detail by Deleuze and Guattari in *A Thousand Plateaus* (1987), and which will be discussed in the following chapter. More important for the task at hand, however, is Kropotkin's recognition that anarchism and statism are not entirely new inventions, but have their own lineage. There have always been revolutionists, he argues, but 'some of them, while rebelling against the authority that oppressed society, in nowise tried to destroy this authority; they simply strove to secure it for themselves' (1912: 3). Over against this, another current can be seen, one that sought 'not to replace any particular authority by another, but to destroy the authority that had grafted itself onto popular institutions, *without creating a new one to take its place*' (3; italics added). Kropotkin thus clearly formulated, for the first time, the struggle between hegemony and affinity, and was therefore able to identify, again for the first time, an anarchist ethics and politics that set out to battle the will to hegemony not only in others, but *in its own theory and practice.*

One way in which this realization plays itself out is visible in Kropotkin's handling of the problem of the transparent society. At the time of the *Conquest of Bread*, he envisages the revolution as an event that will 'put an end to exploitation' (1990/1892: 161), and speaks of a 'tendency of the human race' towards this glorious endpoint (1990/1892: 38). But the non-hegemonic anarchist ethic that Kropotkin was developing was incompatible with a world-encompassing ideology, even one that would be 'freely chosen' by any 'rational' being. In his 1899 *Memoirs of a Revolutionist* he maintained that an anarchist society 'will not be crystallized into certain unchangeable forms, but will continually modify its aspect' (1942/1899: 114); and in a formulation that smacks of deconstruction *avant la lettre* he declared that he 'conceive[d] the structure of society to be something that is never finally constituted' (Kropotkin in Buber 1958: 43). In his later formulations, Kropotkin expected the anarchist-communist world to arrive not in a final, fell swoop, but via a process of *ongoing* expropriation—or structural renewal based on mutual aid. In finding a way out of so many of the impasses that had afflicted anarchist theory since Godwin, and in anticipating many of the insights of poststructuralist theory, Kropotkin appears as a key figure in the genealogy of affinity—the first *post*anarchist to begin to emerge out of the modernist quagmires of eighteenth- and nineteenth-century socialism.

EARLY TWENTIETH-CENTURY ANARCHISM
AND THE CONCEPT OF STRUCTURAL RENEWAL

Kropotkin's many contributions to anarchist political theory, science and culture are of course important in and of themselves, and it would be ridiculous to suggest that I have done justice to him or to any other of the figures I have briefly discussed in this chapter. Rather, I have focused the reader's attention quite fixedly on a genealogy of the logic of affinity, in which Kropotkin's work appears as crucial in preparing the ground for Gustav Landauer, who I would argue went as far as possible, within the constraints of modern revolutionary practice, in effecting a break in the logic of hegemony. Not well known outside of anarchist circles, and a minor figure even within them, Landauer lived, wrote and agitated primarily in Germany in the late nineteenth and early twentieth centuries, and was murdered for taking part in the Bavarian uprising of 1919.[10] In keeping with the anarchist principle of means/ends coherence, but breaking with its long-standing reliance upon 'the people', Landauer insisted, in *For Socialism* (1978/1911), that a radical transformation of state-capitalist societies could not be achieved by the instantaneous destruction of existing institutions, or by their slow reform, or even by some combination of the two. Rather, new institutions must be created 'almost out of nothing, amid chaos' (20); that is *alongside*, rather than inside, existing modes of social organization. He argued that the social revolution should be carried out here and now, for its own sake, by and for those who wished to establish new relationships not mediated by state and corporate forms.

For this strategy the appropriate tactics involve a complementary pairing of disengagement and reconstruction. 'Let us destroy', Landauer suggested, 'mainly by means of the gentle, permanent, and binding reality that we build' (93). To the extent that it does not seek an abrupt and total transition away from capitalist modes of social organization, Landauer's strategy of shares with reformism a willingness to co-exist with its enemies. However, it is crucially different from reformism in that it does not provide positive energy to existing structures and processes in the hope of their amelioration. Rather, it aims to reduce their efficacy and reach by withdrawing energy from them and rendering them redundant. Structural renewal therefore appears simultaneously as a negative force working against the colonization of everyday life by the state and corporations, and as a positive force acting to reverse this process via mutual aid. Just as the states and capitalism advance by percolating into everyday relations, structural renewal proceeds through its own dispersion of regularities, its own viral infections and subtle transformations.

Structural renewal also differs from all previous conceptions of Revolution—including what I have called Kropotkin's theory of (r)evolution—in that the construction of alternatives is not seen as preparatory to a coming event that gives a transcendent meaning to what would otherwise be mere quotidian labour. He states very clearly that 'the transformation of social institutions, of property relations, of the type of economy, cannot come by way of revolution' (21). Here Landauer focuses our attention on the fact that in most modernist Revolutionary theories the building of a new world is curiously devalued; it is as though participating in a revolution is more important than living in the world it is supposed to bring about, as though neither the means nor the end is of much value without a singular passage through an ecstatic phase of creation/destruction. With Landauer, the construction of alternatives finally appears as valuable *in and of itself*, and the revolution (which no longer should be known by this name) appears as an ongoing effort in 'love, work, and silence' (21–2).

In arriving at this position Landauer was heavily influenced by Nietzsche's sociological method and his critique of modernity, so it should not be surprising that many of the central insights of an early twentieth-century anarchist anticipate those of poststructuralist theory. Perhaps the most crucial advance made by Landauer was his recognition that the state is not a 'thing', an instrument to be wielded by a dominant class, as in classical marxist theory, or by the representatives of a pluralistic set of interests, as in postmarxism and (neo)liberalism. Rather, Landauer insisted that the state is a condition, a certain sort of relationship. It is a 'nothing', and has to conceal this nothingness by donning the mantle of nationality and connecting nationality with community—by seeking to become (in an obvious reference to Hegel) 'a spirit and ideal' (43). In analysing the state as a set of relationships among subjects Landauer grasped the key insight of Foucault's governmentality thesis—that we are not governed by 'institutions' apart from ourselves, by a 'state' set over against a 'civil society'. Rather, *we all govern each other* via a complex web of capillary relations of power.

Landauer sees capitalism in the same way, as 'a nothing that is mistaken for a thing' (132), a set of *relations* between human individuals and groups. The same analysis is applied to law, religion, education—all of the sociological institutions are analysed as 'names for force between men [and women]' (132). For Landauer, then, because capitalism, the state—and of course socialism as well—are all modes of human co-existence, changing these macrostructures is very much a matter of changing microrelations: new forms 'become reality only in the act of being realized' (138) and 'revolutionaries come into existence only by means of the revolution' (82). Read in the context of a nascent form of Nietzschean discourse analysis, of a sociology of relations of power, Landauer's 'individualism' and 'voluntarism' appear in a different light from that cast upon them by marxist theory—if the state is in all of us, in how we live our lives, then living without the state form means living our lives differently, as individuals and as members of diverse communities.

As we have seen throughout this chapter, the question of community, of who will make the revolution, has dogged anarchist theory and practice since its inception. Some have advocated educating the masses for their role in the construction of a new world order, others have insisted that they need to be secretly seduced into revolutionary activity, and still others have admitted that the people may in fact have to be forced to be free. None before Landauer, however, had questioned the assumption that the masses, the people—the many, by whatever name—are the necessary agents of any radical social transformation. '[W]e here do not have to go along with the foolish and shameless flattery of the proletariat,' he proclaims, 'since socialism aims at the *abolition* of the proletariat ...' (49; italics added). In his desire to rid socialist theory and practice of an excessive orientation to totalizing social change and, once again, under the influence of Nietzsche, Landauer ventures so far away from faith in the revolutionary credentials of the masses as to land himself in a position that seems painfully lacking in solidarity with those who suffer the most under capitalism. What kind of socialism is this, one might well ask? What of solidarity, affinity and mutual aid?

To understand Landauer's position it is necessary to read beyond these barbs, to appreciate that he is facing, finally, a central dilemma of anarchist theory. We cannot wait for 'everyone' to choose to live in non-statist, non-capitalist relationships, or we will very likely wait forever. Nor can we *force* socialism on anyone, since that would violate our commitment to respecting the autonomy of individuals and groups. Hence there is no choice for those of us who desire to live differently but to begin to do so ourselves. 'That is the task', Landauer observes in one of his better moments: 'not to despair of the people, but also not to wait for the people' (138). Already, in the early 1900s, Landauer was able to observe that there were no hard-and-fast lines between exploiters and exploited, that capitalism was not polarizing the classes in the way that Marx had predicted. 'One ought not to speak of capitalist entrepreneurs under the assumption that the existence of capitalist society depends particularly on their number. Rather one ought to speak of how many have a stake in capitalism, of those who, as regards their external livelihood, enjoy relative prosperity and security under capitalism' (75). While neoliberalism is certainly increasing the distance between rich and poor all over the world, it remains the case that a vast silent majority in the G8 countries perpetuates not only the domination of the planet and the peoples of 'the rest of the world', but their own domination as well. On this score I believe that Jean Baudrillard is quite correct: the revolution has in fact occurred, the masses of the First World have chosen quiescence, and nothing we can do will change their behaviour for the better.[11]

At this point a familiar question emerges: if existing social relations are to be rendered redundant, then what will take their place? Like Kropotkin and Bakunin, Landauer did not offer a vision of a New Harmony. Rather, he always refused to say how a new socialist reality 'should be constituted as a whole' (29), and held that the point of structural renewal was not to 'establish things and institutions in a final form' (130). Rather, he insisted that the building of socialism would require a spirit of creativity and improvisation. 'We need attempts', he argued. 'We need the expedition of a thousand men [*sic*] to Sicily. We need these precious Garibaldi-natures and we need failures upon failures and the tough nature that is frightened by nothing' (62). The theme of the experiment thus returns, but this time in a non-rationalistic form. As I will discuss in the following chapter, it is the kind of experiment associated with nomad rather than Royal science, the experiment conducted by the smith rather than the citizen.

CONCLUSION: TAKING THE TALLY

Indeed, with Landauer we find the full fruition of the anarchist response to Marx and Engels' critique of Utopian socialism. To the charge that Utopian socialism seeks to emancipate 'everyone at once', the anarchists responded by elaborating a non-hegemonic theory of social change that defied the revolution/reform dichotomy, that *did not seek to free anyone at all* but focused on how each of us, as individuals and members of communities, must free ourselves, in an effort that cannot be expected to terminate in a final event of revolution. Objecting as much as the marxists did to rationalistic social experiments, but also rejecting marxist 'planning', the anarchists evolved a theory and practice of non-rationalistic social experiments, an empirically-based search, if you will, of the ever-shifting problem-solution spaces offered by modern western societies. Finally, as I hope to have shown in this chapter, anarchism after Utopian socialism was far from a mish-mash of theories. Bakunin, Kropotkin and even Landauer saw themselves as politically motivated social scientists. Individual intent, however, is a sorry category to invoke in the context of a genealogical analysis. Instead, let us note the existence of a clear line of argumentation and political practice: the emerging theme of structural renewal guided by a logic of affinity, that was always already present in Godwin but had to go through the vicissitudes of a century of interaction with the hegemonic forms of liberalism and marxism in order to find itself, as it were, as a theory of social revolution distinct from both marxist political revolution and liberal political reform.

This is a theory and practice that is decidedly topian rather than Utopian. It is about the here and now, while not pretending there has been no past nor that there will be no future. For, as Kropotkin so elegantly showed, the logic of affinity is ever-present, even in the most advanced forms of (post)industrial bureaucratic control. It is not a dream, but an actuality; not something to be yearned for, but something to be noticed in operation everywhere, at every moment of every day. Thus we can perhaps see that the ongoing battles between Scientific and Utopian socialisms are driven by a narcissism of small differences and a mutual ignorance of the advances made by the opposing tradition. And, as postmarxists and postanarchists have pointed out with respect to their own traditions, the theorists on *both* sides of the classical debate were Utopian in a sense that did not occur to either of them—that is, they all believed in the possibility of a society in which relations of domination and exploitation would be entirely eliminated. In the classical socialist Utopia, be it marxist or anarchist, struggle and relations of power were expected to give way to fishing, philosophy and the apolitical administration of things. Poststructuralist critique has the potential to ward off this danger, but at the risk—as in postmarxism—of a slide into a still hegemonic non-capitalist liberalism.

There are, however, other possibilities, other openings created by the poststructuralist re-reading of classical socialism; possibilities opened up by postmodern anarchisms and marxisms that have abandoned the revolution but retained their commitment to radical social change. These theoretical tendencies help us to grasp the logic of the newest social movements in its specificity, through new ways of conceptualizing the subject of political action, as well as the collectivities in which s/he participates, which s/he in fact *creates and maintains* through her/his activity. Here I am speaking not of the 'citizen' of the state/civil society nexus, nor of the revolutionary/ libertarian nomad, but of the smith, the autonomous subject of the coming communities.

5
... and Now

The genealogy of affinity does not end with the classical anarchists, of course, just as it did not begin with the anti-globalization activists of the late 1990s. Several threads have yet to be examined, all of which are tied together by their deconstructive relation to the modern socialisms of the eighteenth to mid-twentieth centuries. I use this term in its strict Derridean sense, to refer to a way of working self-consciously within a tradition, revaluing its values, questioning its questions, to produce something new, something other. Not an abandonment of the past, not a synthesis or even a progression, but an intimately connected divergence or line of flight. This is the spirit in which I have read some of the key texts of classical anarchism, in search of a theory and practice that breaks with the hegemony of hegemony. There it was possible to find much of what was needed, but not quite everything; certain threads tangled up with the discourse of classical anarchism need to be untangled and set to the side, while others need to be retained in a new weave. Postanarchism, as I noted in the last chapter, has taken on this important task of a creative re-reading and re-writing of the anarchist tradition in response to postmodern social conditions and poststructuralist critique, and has opened up some very intriguing questions. In this chapter I want to explore postanarchism's ethico-political commitments, that is, its conception of community, the political subject and relations between them. I will present a critique of both the purely nomadic subjectivity articulated by writers such as Lewis Call and Hakim Bey, and the moral-Habermasian citizen described by Todd May. As an alternative, I offer a brand of postanarchist subjectivity based on Deleuze and Guattari's concept of the smith—a subject that is neither 'free-floating' nor irredeemably attached to any particular moral imperative.

Postanarchists are not the only writers who have been working towards what might be called a poststructuralist communism.[1] For many years Italian autonomist marxists have been interpreting and further developing the work of Foucault and Deleuze, and have been particularly fruitful in developing concepts appropriate to the analysis of the societies of control. In their analysis of relations between state and community forms, Landauer and Kropotkin were remarking on the process that Michael Hardt and Antonio Negri have recently characterized as the real subsumption of civil society by the state and capital (1994),[2] and it could be said that the distinction between social and political revolution has an analogue in autonomist marxist theory in the conceptual pairing of constituent and constituted power. I am not arguing, of course, that Hardt and Negri are coming out of the closet wrapped in black flags. Nor do I make the mistake of thinking that Todd May is a marxist—or even an anarchist in the political sense, for that matter.[3] Rather, I would argue that some autonomists—the ones who have most fully left behind their leninist baggage—share with some anarchists—those who have, for their part, abandoned the dream of a transparent society—are following a common trajectory in their attempts to break with the hegemony of hegemony. Over the following two chapters a similar argument will be made with respect to Native American political theory, certain postcolonial and postmodern feminists, and Judith Butler's queering of the NSM-based theory of political identity formation. The goal is not to reduce any of these disparate paradigms to one another, or to a common denominator among them all, but to trace the logic of affinity as a regularity dispersed, in different ways and at different times, amidst and between them.

In addition to maintaining the focus on the genealogy of affinity, I also want to bring the reader's attention to a key question for those who advocate and struggle for radical social change: What is it about the status quo that renders it, no matter how horrible, almost impervious to transformation? As shown in the previous chapter, anarchists have for a very long time struggled with the apparent necessity of putting their faith in 'the people', just as marxists have fought their own battles to overcome 'false consciousness'. It has become apparent to some tendencies within both traditions that action and struggle over long periods of time are necessary to transform subjects who enjoy giving away their autonomy into subjects who are willing to take on the work necessary to preserve it. This issue has also been taken up by poststructuralist theorists like Foucault, Butler, Deleuze and Guattari, who have helped us to understand further how and why it is that subjects desire their own repression. Unfortunately, we get little out of Foucault in terms of concrete strategies and tactics for community-based activism, and while Deleuze and Guattari are less reticent on the question of radical social change, much of what they have to say is obscurely poetic and jargon-laden. Finally,

Butler, while more 'practical', seems to have retreated into the kind of postmarxist identity politics associated with Laclau and Mouffe, which I have argued reproduces the hegemony of hegemony and the politics of demand (see her contributions to Butler, Laclau and Žižek 2000). Thus the question remains: how can we work to create more opportunities for more people to choose a life of autonomy over one of subservience? How can we provide those who do take a line of flight from the state, capital, heterosexism, racism and the domination of nature with more places to land, with *other* places to land? To begin to answer these questions, I want to recall briefly some of the well-known aspects of the poststructuralist critique of capitalist modernity and provide a more in-depth discussion of some themes that have been given less attention.

ELEMENTS OF POSTSTRUCTURALIST CRITIQUE: BECOMING MINOR

Although it is now widely acknowledged that poststructuralism is at best a label of convenience, it is also obvious that certain common themes *are* dispersed in the texts of an extremely productive and influential generation (or two) of French writers. This nexus would include figures such as Jacques Derrida, Julia Kristeva, Hélène Cixous, Jean Baudrillard, Michel Foucault, Gilles Deleuze, Félix Guattari and Jean-François Lyotard. In the English-speaking world—especially among those who are not comfortable with the implications of poststructuralist critique—some or all of these thinkers are often seen as 'postmodernists', that is, as rejecting outright the basic tenets of European Enlightenment. This label has been applied to Derrida and Foucault for example, despite their repeated protestations in interviews and textual demonstrations of their indebtedness to modern theory and philosophy. In 'What is Enlightenment?', Foucault declared his commitment to a 'critical ontology of ourselves' which he associated with Enlightenment critique (Foucault 1997a: 319), and Derrida has explicitly situated deconstruction as a critical engagement with marxism (Derrida 1994). Calling Baudrillard and Lyotard postmodernists makes a little more sense, since they have in fact identified themselves with both the sociological claim that a postmodern condition exists as a set of currents running beyond/ against European modernity, and have to some extent happily (and of course ironically) taken on the mantle of the 'anti-theorist'. My own belief is that nothing can obviate the need to read particular writers closely and carefully on their own terms, and to be attentive to the disparities that exist between those who are supposedly part of the 'school' of poststructuralism; to pay more attention to texts, that is, than to categories. For this reason, I deploy the term 'poststructuralist critique' only as an admittedly inadequate intellectual convenience, knowing full well that it elides many crucial divergences among the work of those whom it seeks to encompass.

On this basis, it is by now not controversial to note that Derridean deconstruction shares with Foucauldian genealogy an orientation to challenging received dichotomies and the relations of power with which they are associated. Along with the other writers mentioned above, Derrida and Foucault also share a scepticism about the ability of the Enlightened European subject to know itself and the world in an objectively correct fashion, and are dubious of any claim that there is a universal flow of history towards some future moment when relations of power as domination will be entirely eliminated. Most of those who are included in the poststructuralist canon were marxists of some kind at some point; most took a distance from marxism in the 1950s or 1960s, and most never fully returned to what appeared to them to be a dominating response to domination. In the case of Kristeva and Cixous, this critique of the modern European humanist subject was extended to cover its implicit masculinity and, as I have noted previously, has been utilized by postcolonial and queer theorists to challenge its racialization and its heteronormative bias. The result of all of this work was that, sometime in the late 1980s, critical social, political and cultural theory reached a point where everything had been deconstructed, that is, where the contingency of all identities and relations had been exposed.

At this point, some have claimed, there began a 'return of the political', that is, a return of questions related to *changing* dominant institutions and identities, rather than merely exposing their contingency (Mouffe 1993). This thesis has been adequately refuted, I would say, by writers such as Simon Critchley, who has convincingly argued that Derridean deconstruction has *always been* ethical and political; that it makes no sense whatsoever outside of its commitment to certain values and political projects over others (Critchley 1999). I would suggest that the same argument can be made with respect to the other so-called poststructuralists/postmodernists, including even Jean Baudrillard, whose work is driven by an intense undertone of ironical critique of the societies of simulation and control, and whose dismissal of sociology is deeply sociological in the content of its insights and the form of its expression. At any rate, the claim that the political has 'returned' does have the merit of focusing more attention upon the value-orientations of poststructuralist critique, and makes it possible to pose the question of poststructuralism and the political a little differently—it allows us to ask, not why those elitist theorists refused to say anything about politics, but why so many people have failed to understand what they did have to say as being political. By now the answer should be clear—poststructuralist politics contains vital elements that are non-hegemonic, and are thus impossible to comprehend within the liberal-marxist paradigms of state-based social change.

The fact that non-hegemonic elements are central to poststructuralism does not mean, however, that this discourse cannot be appropriated for hegemonic ends. In Chapter 3 I discussed what is perhaps the best-known mode of post-politics, the 'postmarxism' of Laclau and Mouffe, Lefort et al. The central insight of this paradigm is that 'power is an empty place' (Lefort 1988: 232–3), that is, that in postmodern societies there is no particular person (say, the king), nor any particular institution (say, the state) that can be seen as the sole locus or fount of relations of domination. Rather, power is seen as disseminated through many relationships, every day and every night, personal and political, discursive and material. In such a context political revolution makes no sense, as there is no building one could seize, no leader one could assassinate, in order to eliminate power effects and achieve a transparent society. In dispensing with the revolution, postmarxism takes an important step away from Enlightenment socialism. The problem, however, is that it does so only to land squarely in Enlightenment liberalism, thereby effecting more of a return than an advance. While I would not deny that postmarxist liberalism represents one way of drawing political implications from poststructuralist theory, I am much more interested in readings that pay closer attention to the radical critique of capitalism and the generalized state form that permeates poststructuralist texts, as well as the re-imaginings of community, science and politics that go along with this critique. I am much more interested, that is, in the postcommunist moments of poststructuralist critique than in its liberal leanings. 'Forget capitalism and socialism', Guattari and Negri declared in 1985; the project is to 'rescue "communism" from its own disrepute' (1990: 7).

What, then, are the elements out of which a non-hegemonic poststructuralist politics might be constructed? First, I want to note that the observation that power does not emanate from a single point, but flows through all points, does not necessarily lead to a liberal position that accepts the state form as a 'necessary evil'. That all modes of social organization involve relations of power does not mean that particular modes cannot be evaluated according to the extent to which they *encourage* or *discourage* the maintenance, emergence and development of equitable relations between autonomous individuals and groups. I have argued that the postmodern liberal-capitalist system of nation-states does not do this. Rather, it does quite the opposite through systems of integration supported by a politics of demand. The question should be: how can relations between human individuals and groups and the natural world be structured so as to *minimize* domination and exploitation, taking the *entire* social/natural field into consideration in the *long term*? Can we avoid, for example, the kind of circular silliness involved in using the coercive power of the state (through laws and policies) to limit the damage done by capitalism, while at the same time hoping that capitalism (through free markets) will ameliorate the effects of state domination? This is the kind of question asked by a radical poststructuralist politics that maintains a thoroughgoing analysis and critique of capitalism, the state form, and all of the interlocking modes of oppression that have evolved along with these two great focii of modern revolutionary activity. In formulating questions in this way, poststructuralism owes much to marxism, and has acknowledged this debt. There is also, as I will show, a more subterranean reliance upon themes derived from the anarchist tradition, for example in Foucault's analyses of biopower and governmentality, Deleuze and Guattari's genealogy of the state form, and Deleuze and Foucault's observations on the societies of control.

The concepts of governmentality, biopower and society of control all describe ways in which the state form has been generalized, or disseminated, out of the 'public' realm assigned to it in liberal theory and into the supposedly inviolate world of the 'private' lives of its citizens. In his work on the emergence of the prison (1979), Michel Foucault argued that prior to the 1700s, those who violated social norms in the European countries were seen to have offended God and the King, and were subjected to punishment by torture and/or execution. That is, after the fact of a supposed offence, physical force was used in a public display to inflict pain upon the body of the offender, through simple machines such as the rack, red-hot pincers, or drawing and quartering. With the rise of Enlightenment humanism and republicanism, bodily interventions such as public punishment/execution came to be seen as excessive and perhaps even 'uncivilized'—certainly not in keeping with the new dignity of the 'individual'. The dominant mode of handling exceptional behaviour thus shifted towards interventions carried out in enclosed spaces—the prison was born. One offended now against the state and its laws rather than against a god-king, and although the body was still affected by imprisonment, the goal was to affect what Foucault calls 'the soul'; to change the person or, more precisely, to cause him to want to change him- or herself.

In the prison one was subjected to constant scrutiny via systems such as the panopticon, which allowed a small number of guards to observe the activities of a large number of inmates (1979: 201). This was a key innovation for Foucault, as it marks a hinge point between the disciplinary regime of the prison and the technologies of biopower that began to arise in the 1800s. In *The History of Sexuality* (1990) and his lecture on 'Governmentality' (1991/1978), Foucault describes how attention was shifted from the individual to the population, from self-regulation to social regulation, from crime to deviance. One no longer offended the state or the king, one offended the social order; and in a sense, social control went public again, as the technologies that were previously reserved for use in prisons and other enclosures began to be deployed beyond their walls. This shift to a generalized surveillance/prediction mechanism, of a vast dispersal of state coercion, is what Foucault calls governmentality or biopower. In such a system it no longer makes sense to speak of 'the state' as a locus of relations of domination, since relations of domination are everywhere. It no longer makes sense to speak of 'the king', since kings are now found in families, convents, factories and schools. *We all become agents of social regulation, we all watch each other*; the state becomes, as Landauer suggested, a state of relationships. And it is not surprising that the population should go along with this shift, for it was in keeping with the highest ideals of the European Enlightenment—responsible individuals, free to make their own decisions, were contributing to what was now defined as the ultimate end of government: 'the welfare of the population, the improvement of its condition, the increase of its wealth, longevity, health etc.' (Foucault 1991/1978: 100).

Biopower obviously remains with us today, but so do discipline and punishment. Prisons are alive and well, and people are being tortured and executed every day, all over the world. It is important to remember that in identifying these shifts, Foucault was working genealogically, pointing out the emergence of new forms out of old, leading inevitably to intermingling and intermixing—there is no such thing as a clean break. But each extant form shifts as the dominant mode shifts, and biopower is no exception. In the mid-1900s, systems of genetic and electronic engineering began to appear and quickly rose to ascendancy in the western world. These new tools of cybernetic rather than social regulation made it possible to increase the efficacy of surveillance and management, leading to the advent of what Deleuze, observing a trend in Foucault's latest thought, identified as the societies of control (Deleuze 1992). In control societies, the focus of management efforts is on neither the individual body nor the population as a whole, but on the acquisition of power over life itself, power to 'make live and to let die' (Foucault 2003/1976: 241). In such systems one does not need to worry about offending a god-king, a state or even social norms; rather, one must avoid giving rise to an exception within the cybernetic system, one must not become, quite literally, an 'error'.

The societies of control put all of us in an extremely ambivalent position. Since they make it increasingly difficult to do anything without the appropriate password, card, clearance, number, it is also difficult to 'do anything wrong'—that's the point of societies of control, not to react to what we do, but *to make it impossible* for us to do anything that is not optimized for state control and beneficial for capitalist exploitation. At the same time, however, *everything* we do becomes a problem, or at least a potential problem, due to the intensity, ubiquity and fallibility of the systems deployed to keep us in line. No human being can live up to the standards of perfection they demand, so we are perpetually failing to make payments on time, filling out forms incorrectly, taking our medication in the wrong dose, and so on. We are always already enveloped in error, always already deviant, problematic and therefore deserving of the very imprisonment that has put us into this situation in the first place. Taken to its ultimate end, the society of control becomes *The Matrix*, a perfect virtual world of total envelopment and control of utterly docile bodies.

With this genealogy—roughly sketched and terribly compacted here—Foucault contributes to the project inaugurated by Kropotkin in *Mutual Aid* and carried on by Landauer, by showing how autonomous communities and individuals have been seduced/coerced into statist relationships, into living a life based on relations of power as domination. Although he clearly did not identify as an anarchist in the political sense, Foucault did at one point present his conception of power, upon which his analysis of the state form is based, as explicitly anarchistic:

> I am not saying that all forms of power are unacceptable but that no power is necessarily acceptable or unacceptable. This is anarchism. But since anarchism is not acceptable these days, I will call it anarcheology, the method that takes no power as necessarily acceptable. (1980)

In taking up this position, Foucault both locates his project within and takes it beyond classical anarchist theories of the state. It is within the anarchist tradition in the sense that he sees life without the state form as an *ongoing actuality* rather than an impossibility (as in liberalism), or as a utopian point to be reached in some far-off future (as in most marxisms). What takes it beyond classical anarchism is his disavowal of the possibility of living a life *entirely without* relations of power as domination. Rather, Foucault sees that within each of us as individuals, and within any group, there is a potential for things to go either way, or to go both ways at once. Just as there is no pure freedom, there is no pure domination.

This insight also animates the work of Deleuze and Guattari who have, like Foucault, made important advances in the analysis of the state form. I cannot possibly do justice to the complexities of their work here, but none the less I would like to outline some of its key elements as they relate to the discussion at hand. First, it must be understood that Deleuze and Guattari are much more interested in what they call the state form than they are in any particular state. That is, their analysis operates at a high level of abstraction, seeking regularities across a disparate array of actually existing states. It could, therefore, be seen as idealist and/or essentialist, and perhaps should be seen in this way. But I hope to show how the notion of the state form can be of use in the practical task of better understanding, and thereby more successfully avoiding, statist relationships. As understood by Deleuze and Guattari, the state form is as an 'apparatus of capture', a system that conditions its surroundings so as to perpetuate and enhance its own existence, by bringing 'the outside' to 'the inside'. They suggest that this propensity is 'magical', in the sense that 'it always appears as preaccomplished and self-presupposing' (1987: 427). The liberal theory of the state, for example, casts our submission to the Leviathan into the dim reaches of a pre-civilized era, at the same time as it uses this myth to ensure our compliance in the present. Currently, we can see how the neoliberal state attempts the capture of relationships previously outside its grasp through their submission to the 'inevitable' trends of globalization in all of its forms, such as the supposedly free mixing of races and ethnicities, supposedly free trade, and so on. What I have called integration can be seen as the mode of capture appropriate to both the postmodern nation(s)-states (such as Canada and the EU) as well as the emerging world-state–capitalist system.

But this is not to say that the state form is modern in its genesis, or that it has a genesis at all. Rather, Deleuze and Guattari take issue with Marx and Engels' contention that the state arose through the Imperial capture of the surplus generated by sedentary agricultural communities. 'On the contrary,' they argue, 'the State is established directly in a milieu of hunter-gatherers having no prior agriculture or metallurgy, and it is the State that creates agriculture, animal raising, and metallurgy; it does so first on its own soil, then imposes them on the surrounding world' (428–9). As is often the case in their work together, Deleuze and Guattari get carried away with their polemic so that, in challenging the received version of the 'origin' of the state, they begin with a simple inversion—the state is said to 'come first' and produce agriculture and metallurgy, rather than the other way around. However, they immediately complicate this position by suggesting that the emergence of *all* of these forms is 'like seeds in a sack: it all begins with a chance intermixing' (429). In such circumstances one cannot assign cause and effect; one can only note that agriculture, metallurgy and the state form all arose together.

But Deleuze and Guattari want to say more than this. They want to claim that 'there have been States always and everywhere' (429), that is, that the state form was not 'invented', nor did it 'emerge', but exists as an 'immemorial *Urstaat*' (427), a way in which human relations at any given time and place *might* be organized. In the passages I am citing, they even refer at one point to capture as the 'interior essence or unity' of all states (427). To many this will seem to be a move that is uncharacteristic of poststructuralist theory, which, after all, is supposed to be committed to anti-essentialist modes of analysis. But Deleuze and Guattari establish that the state form 'comes into the world fully formed and rises up in a single stroke' (427) only to place it alongside its radical yet complementary other, which they call the war machine. The war machine is that which is exterior to the state apparatus, that which has not been captured and resists capture. Here the metaphor of roving bands or packs is deployed, reminding us of the way in which the nomad appears in an archetypal nightmare of European civilization—galloping in off the steppes, sweeping away everything that matters: houses, walls, fields, institutions, lives. That the East has fared no better is obvious in China's monument to state insecurity, the Great Wall. Thus we might say that what the state form tries to do, the war machine tries to undo. Indeed, Deleuze and Guattari suggest that the best way to avoid the formation of a state is through war—an observation that makes sociological sense when we remember Max Weber's famous dictum that the state is that entity which seeks a monopoly on the use of force in a given territory. Break that monopoly and you break the state, or at least that particular state.

But this relation of exteriority to the war machine does not exhaust the possibilities of the state form. In keeping with their tendency first to posit, then to split 'essences', Deleuze and Guattari are careful to show how these two modes exist in relation to one another. The state not only can, but *must* capture a war machine for itself, it must have its own warriors, which it turns into soldiers. Working for a particular state, a 'tamed' war machine captures subjects and objects and destroys *other* states, so that its components become available for integration. Once again, examples from current world affairs are easy to find. The soldiers sent by the western powers to Afghanistan and Iraq are carrying out the task of suppressing/destroying local war machines ('warlords'/'tribal chieftains') and forms of community ('Islam') that resist integration into the global capitalist system on the terms preferred by the western powers. Without the realizable threat of violent death (that is, without a captive war machine), the United States would not be able to bring 'peace' to anyone, and thus would not be able to fulfil its role as a central node of state/capitalist power in the neoliberal world order. Thus a paradox: the war machine and the state form are always at odds *and* always intermixing: 'it is not in terms of independence, but of coexistence and competition in a perpetual field of interaction, that we must conceive of exteriority and interiority, war machines of metamorphosis and State apparatuses of identity, bands and kingdoms, megamachines and empires' (360–1).

Let us think, once again, about Kropotkin's thesis in *Mutual Aid*. He argues that 'the States, when they were called later into existence, simply took possession, in the interest of the minorities, of all the judicial, economical, and administrative functions which the village community already had exercised in the interest of all' (1989/1902:

151–2). Kropotkin differs from Deleuze and Guattari in positing the emergence of the state form at some defined point in history; that is, he did not fully grasp, as did Landauer, that the state is a state of relations. Put another way, we could say that while Kropotkin recognized that the logic of affinity was ever-present as a potential way of structuring human relationships, he did not see the logic of hegemony in this way. Rather, he saw it as something that could, and should, be tossed into the dustbin of history. This assumption, which arises from the remnants of his faith in the revolution, has important strategic consequences. It obscures a critical danger that lies in wait for all who attempt to live a non-statist life, the danger that the state form will return, just when we thought we had got rid of it for good. This is not only a problem when constructing socialism in one country—or region, or town—but would remain even in a world entirely devoid of states. This is because while we might rid ourselves of particular states, *we can never rid ourselves of the state form*. It is always already with us, and so must be consistently and carefully warded off.

This problem—which was not addressed by the classical anarchists—is handled in an interesting way by Deleuze and Guattari. Citing Pierre Clastres' argument in *Society Against the State* (1989), they note that so-called 'primitive' societies maintain 'collective mechanisms of inhibition' that prevent the emergence of hegemonic relationships (Deleuze and Guattari 1987: 358). One of these is modes of leadership that do 'not act to promote the strongest but rather inhibit the installation of stable powers, in favour of a fabric of immanent relations' (358). This argument, of course, is a common trope within anarchist anthropology, from Kropotkin's *Mutual Aid* to Harold Barclay's *People Without Government* (1992), and by adopting it Deleuze and Guattari take on an obvious, but unacknowledged debt. But they do so at the risk of picking up a problem inherent to all modernist anthropologies, namely a tendency to see societies without the state as representative of a stage that necessarily comes before, and leads up to, the higher forms of 'civilized', or state-based, societies. Despite some lapses of language that seem to indicate they are not up to the task of overcoming primitivism—such as the apparently unqualified use of the term 'primitive' itself—Deleuze and Guattari do in fact present a critique of Clastres on just this basis:

> He [Clastres] tended to make primitive societies hypostases, self-sufficient entities ... he made their formal exteriority into a real independence. Thus he remained an evolutionist, and posited a state of nature. (1987: 359)

Despite the many points at which their writing remains Eurocentric, the authors of *A Thousand Plateaus* were able to take at least one crucial step around the trap of evolutionary thinking by insisting that 'bands and clans are no less organized than empire-kingdoms' (359). Only from the point of view of a triumphant and self-congratulatory modernity do non-statist societies appear homogeneous and lacking in structure. On a closer view, it becomes apparent that everywhere there are human beings (or any other entities for that matter) there is social organization.

Once we accept that actually existing non-statist societies are organized societies rather than reversions to, or remnants of, a nasty and brutish 'state of nature', we can begin to see that their traditions offer alternatives to the hegemony of hegemony. This theme will be developed further in the following chapter, but for now I want to note that whereas the argument that war is the most potent antidote to the formation of states seems to lead to a brand of right-libertarianism at best, or to the status quo capitalist-patriarchal free-for-all at worst, this mode of warding off the state form involves indigenous/anarcha-feminist values of community, care and mutual respect. But still the function of 'war' must remain, if by 'war' we mean the willingness, on the part of individuals and communities, to defend themselves rather than be captured. In a gesture reminiscent of the long-standing love/hate relationship between anarchist activists and 'the people', Deleuze and Guattari observe that 'the State apparatus needs, at its summit as at its base, predisabled people, preexisting amputees, the still-born, the congenitally infirm, the one-eyed and one-armed' (426). The ableist language deployed here must be noted and critiqued, but the metaphorical resonances are clear and important: states require subjects who desire not only to repress others, *but also desire their own repression*, subjects who are willing to trade away their autonomy for the promise of security. Warding off the state, then, means primarily enabling and empowering individuals and communities; it means, as Landauer argued, *rendering the state redundant*, and watching very closely that it remains that way. Certainly the men and women of the state understand this requirement from the other side, as evidenced by their systematic attacks on non-statist societies via colonialism, and their incessant work, via secret police forces, the mass media, state schools and 'social security', to ensure that as many subjects as possible remain as dependent as possible; that we continue to beg for scraps of food, land, recognition and tolerance, to be *allowed* to 'let live'.

Although it is important to understand how the state form as such operates, it is impossible to analyse adequately actually existing states without paying attention to their longstanding interactions with other apparatuses. While it is a commonplace to suggest that nation-states are less important now than they were in some era 'prior' to capitalist globalization, it is clear that the state form and capital have grown up together, in a relationship that while it may be fraught with localized and short-term animosities, has been in the long term mutually beneficial. Following the same anti-evolutionary logic that underlies Deleuze and Guattari's concept of the state form, capitalism—as an apparatus of *exploitation*—must also be seen as an ever-present potential, a way of being with others that is always possible, if we desire it—perhaps unconsciously—and must always be warded off if we do not. Just as the state and the war machine intermix, capitalism is in a perpetual field of interaction with socialism, that is, with the idea that exploitation should be eliminated, or at least minimized. And we might say that racism, patriarchy, heterosexism, ableism, the domination of nature and any other discourse that carves up the social-natural field into a hierarchy of identities, are apparatuses of *division* that undermine community-solidarity and thereby facilitate capture-exploitation.

Deployed and extended in this way, Deleuze and Guattari's analysis of the state form shows the similarities between their approach and Foucault's concept of apparatus or *dispositif*.[4] Not only do they share an approach to the analysis of the apparatuses of large-scale social organization. There are also deep affinities in the ways in which they thought alternatives to these apparatuses might be constructed. For this generation, state communism had clearly shown its limits, and the wild ride of May 1968 had culminated in something worse than a return to the status quo, since it seemed as though not only this particular revolution, but the revolution as such, had made its final exit from European history. The generalization of the state form into all relationships and the ubiquity of the electronic technologies associated with globalization and the societies of control seemed to rule out any desire for, and therefore any possibility of, radical social change on a mass level. These thinkers and activists of the European centre could have turned their attention to the so-called 'periphery', and to some extent they did. But only in a glancing way, only with a tell-tale lack of the kind of close attention they paid to their readings of Nietzsche and Heidegger. Spivak was right: at times it really did appear as though Deleuze and Foucault thought their only responsibility was to transparently allow the subaltern to speak.

But this is perhaps to put too narrow an interpretation on the possibilities of the process that Deleuze and Guattari called _becoming minor_. As Rosi Braidotti has pointed out, this is commonly thought to involve a kind of Kerouackian 'narcissistic self-glorification' (1997: 68), something suburban White men do in their twenties before settling down and becoming stockbrokers or university professors.[5] On this reading, becoming-minor would involve not only lack of care for others, but lack of care for the self as well. Braidotti, however, sees this process as 'life on the edge, but not over it; as excessive, but not in the sacrificial sense (exit Bataille). It is definitely anti-humanistic, but deeply compassionate in so far as it begins with the recognition of one's limitations as the necessary counterpart of one's forces or intensities' (68). Forces and intensities, Braidotti adds, necessarily involve interaction with others, and therefore ethical and political commitments—particularly a commitment to a 'space of becoming ... posited as a space of affinity and symbiosis between adjacent forces' (69). Thus the minority appears as 'the dynamic or intensive principle of change' in Deleuze and Guattari's work (68), where change is explicitly oriented to avoiding becoming major, is in fact defined in such a way as to associate it inseparably with non-hegemonic practices. This is the space of Guattari's molecular revolution, Foucault's micropolitics, a particularly poststructuralist politics of affinity that has been picked up by theorists and activists working within and across a number of different traditions. Although they have remained to a great extent disparate, these endeavours pose a common question that is both current and ancient: how can a micropolitics simultaneously be a communal politics? That is, how can we organize ourselves so as to minimize domination and exploitation, particularly in a world increasingly colonized by neoliberal globalization and the societies of control? Autonomist marxists and postanarchists each offer their own answers to this question, which I will now set out to discuss and critique.

AUTONOMIST MARXISM AND THE CONSTITUENT POWER OF THE MULTITUDE

Although it can trace a lineage back to at least the 1940s,[6] autonomist marxism remained relatively unknown in the English-speaking world until the publication of Michael Hardt and Antonio Negri's _Empire_ (2000). Because this is such a wide-ranging text, I will not attempt to discuss all, or even most, of its arguments. Instead, I will focus on Hardt and Negri's attempt to theorize radical struggles using the concept of 'the constituent power of the multitude'. Understanding the multitude, of course, requires an understanding of what it is working against—Empire, or the neoliberal world order, which Hardt and Negri analyse using concepts drawn from poststructuralist theory, and from what some have begun to call the 'autonomist school of communication' (Brophy and Coté 2003; Coté 2003). Like the state form and the war machine, Hardt and Negri see Empire and multitude as locked in a symbiotic struggle, taking place not only on the shop floor but in everyday life and throughout the increasingly ubiquitous networks of electronic communication. Although I do not agree with their suggestions regarding strategies for radical social change today, there is little doubt that Hardt and Negri's book presents one of the most theoretically compelling accounts of the neoliberal world order, and provides a basis for further work in a number of areas.

Perhaps the greatest strength of *Empire* is its use of autonomist theory to adapt modern marxist and anarchist categories to an emerging, postmodern social condition. The notion of the 'social factory' is a case in point. As Steve Wright notes, Georg Lukács had argued in *History and Class Consciousness* (1971) that the capitalist factory contained 'in concentrated form the whole structure of capitalist society'. Thus, it could be expected that 'the fate of the worker' would become 'the fate of society as a whole' as industrialization advanced (Lukács in Wright 2002: 37). Mario Tronti picked up on this idea, suggesting in 1971 that eventually 'the whole of society exists as a function of the factory and the factory extends its exclusive domination over the whole of society' (Tronti in Wright 2002: 38). As this process continues, it becomes increasingly difficult to distinguish the factory from the non-factory—its disciplinary pattern, as Foucault would say, is dispersed and delocalized, just as the discipline of the state is dispersed through governmentality. The rise of the social factory can also be seen as a variant of Kropotkin's colonization thesis; just as the state takes over (captures) existing social relations and puts them to work in the name of its own authority, so the factory uses electronic technologies to extend its reach ever further into what used to be 'private' and 'public' (that is, non-capitalist) spaces and times—it 'mines' those relationships and activities from which it can extract what appears to be an adequate amount of profit.

It should also be noted that the concept of the social factory allowed the early autonomists to include non-factory employees in their conception of 'the working class'; it allowed them to begin to see beyond class reductionism and to critique, for example, sexist assumptions about what constitutes 'work'. The classic text here is *The Power of Women and the Subversion of the Community* (1973), in which Maria Rosa, Dalla Costa and Selma James argued that the unwaged labour of women in the home was necessary for the social reproduction of waged labour in the factory. This road could have led to the kind of analyses and politics associated with the new social movements but it was—and remains—a difficult journey for most autonomists to undertake. Like their distant cousins in Critical Theory, they are troubled by what they see as a generalized identity politics associated with 'the poisonous culture of the 1980s, what some might call the culture of postmodernism ... that is dominated affectively by fear and resignation and politically by cynicism and opportunism' (Hardt 1996: 8). In the social factory *everyone* becomes a worker, which deprivileges the point of material production; but at the same time, everyone becomes a *worker*, which reimposes the privilege of an expanded, but still delimited, conception of the working class as *the* identity behind all identifications.

This expansive but integrating conception of the working class is visible in the autonomist categories of *immaterial labour* and *general intellect*. Maurizio Lazzarato defines immaterial labour as 'the labour that produces the informational and cultural content of the commodity' (1996: 133). This labour can be further broken down into two components. With regard to the informational content of the commodity, labour under post-Fordist[7] conditions increasingly involves computerized control systems used in direct production, as well as in the communication functions that are essential to the maintenance of a global division of labour. There is a great deal of immaterial labour involved, then, in being a human resources manager for a multinational travel agency, just as there is in working in a 'special economic zone' to produce the chips for the computers upon which the travel company depends. The cultural content of the commodity is said to arise through the immaterial labour of subjects who are not normally seen as 'workers'—those who 'define and fix cultural and artistic standards, fashions, tastes, consumer norms, and, more strategically, public opinion' (133). Lazzarato notes that while the setting of cultural standards was once 'the privileged domain of the bourgeoisie and its children', it has now become the prerogative of what he calls a 'mass intellectuality'. In a move reminiscent of Gramsci's analysis of the intellectual component of apparently non-intellectual activities, Lazzarato claims that the division between mental and manual labour is becoming less stark, that there is more 'brain' in manual production, and less autonomy, or more 'brawn', in those sectors that are commonly associated with mental work. From my own experience as an academic, I can certainly attest to the many ways in which teaching and research are turning into assembly-line activities: hundreds of students are crammed into lecture theatres where they are force-fed 'points' from a computerized display, while the very topics that we research are increasingly circumscribed by local and national prerogatives based on 'social relevance' defined as state policy applicability or potential for commercialization. Undoubtedly, state bureaucrats and capitalist managers now have as much or more to say about what artists and intellectuals do and how they do it, at the same time as every job seems to require some familiarity with a computer or with computerized equipment.

At the same time as it highlights certain commonalities among workers today, however, the concept of immaterial labour also suffers from a tendency to flatten out a mountainous field of difference. It is clear that the *maquiladora* worker in Mexico and the chip designer in California both partake of immaterial labour. But while the latter is among the highest-paid employees in the world, works in pristine conditions, and either has or doesn't need union protection, the former is subject to dehumanizing conditions both physically and emotionally and faces death or dismissal if she tries to change her situation. When we think beyond the G8 countries, and especially when we think about relations *between* G8 countries and those that supply them with cheap labour, the claim that the division between mental/material labour is being blurred also appears suspect. The assembly-line worker in the 'special economic zone' never gets to design anything, but merely implements the commands of the engineer, who for his part does no menial labour of any sort. It seems that the globalizing information economy is in fact precisely designed to *reinforce* and *exploit* the mental/manual division of labour, in a sense to export the worst effects of capitalist alienation and immiseration to people of colour—and especially women of colour—living outside the walls of fortress G8. All of this to say that while the concept of immaterial labour clearly has analytic value, it needs to be subjected to feminist/postcolonial critique.

With this caveat in mind, I want to explore the links between immaterial labour/mass intellectuality and the concept of general intellect. According to Nick Dyer-Witheford, mass intellectuality should be seen as 'the subjective component' (1999: 222) of general intellect, as the 'variable', or human side of 'a labour of networks and communicative discourse' (Vincent in Dyer-Witheford 1999: 227). But human beings cannot carry out their functions within this system without the help of its 'objective, fixed, machine side' (227), that is, the computer networks and modes of organization of work that are characteristic of multinational corporations and state bureaucracies. Thus Hardt and Negri explicitly link general intellect, biopower and the societies of control, arguing that Empire establishes a 'new relationship between production and life' (2000: 365). They claim to take previous work on general intellect a step further, by considering its embodied, experiential aspects, instead of focusing 'almost exclusively on the horizon of language and communication' (29). With this observation, it becomes possible, or perhaps necessary, to consider bodies, symbolic forms, and mechanical/informational systems—our entire being as postmodern subjects—as intimately enmeshed in a 'vast machine that dominates society' (Negri in Dyer-Witheford 1999: 227).

In a sense this is not an entirely new argument; it may be read, for example, as an updated version of Herbert Marcuse's analysis in *One Dimensional Man* (1964). But autonomist theory has refused to wallow in the kind of quietistic cynicism that is characteristic of the Frankfurt School, insisting instead that while general intellect might be immersed in the biopolitical systems of control upon which Empire relies, it also has the potential to undermine these very systems. The examples that can be cited here are legion, from the adroit manipulation of image-hungry television news outlets by early Greenpeace activists, to the use of email and the web to link activists of all stripes, most commonly evoked in the proliferation of the Zapatista's so-called 'netwar' (Cleaver 1998). Thus, the autonomists argue, general intellect not only enslaves us, but also offers the tools of our liberation. Capitalism, they claim, is once again producing its own gravediggers, but this time they are going to use keyboards rather than shovels.

Indeed, the defining characteristics of autonomist theory is its insistence that it is the workers (in the factories, homes and seaside cottages equipped with satellite internet connections) who have created and sustained capitalism/Empire, not only by allowing their productivity to be captured and exploited, as in the standard marxist analysis, but also through their efforts to 'rupture this recuperative movement, unspring the dialectical spiral, and speed the circulation of struggles until they attain an escape velocity in which labour tears itself away from incorporation within capital' (Dyer-Witheford 1999: 68). In an argument that has strong resonances with regulation theory, the autonomists propose that capitalist systems are prodded to change by working-class struggles; they get bigger, stronger, more resilient, as they respond to each new challenge. The rise of the welfare state as a means of warding off socialist revolution would be an excellent example of the effects of what the autonomists call *auto-* or *self-valorization*. But does the welfare state represent a rupture, an escape? Yes, in the sense that it gives some solace from the worst effects of capitalist immiseration. No, in the sense that it does so only through coercive integration into rational-bureaucratic apparatuses. We escape the manipulative hands of capital, only to fall into the coercive arms of the state. As indicated in the quote from Dyer-Witheford above, there must be more to self-valorization than mere reform. There must be a desire for a rupture, a break with the neoliberal order that, after Deleuze and Guattari, we cannot see as final—if anyone achieves escape velocity, they will eventually be brought back to earth, and so had best be concerned from the outset about where they will land, lest they unwittingly reproduce or be re-integrated by the system of states and corporations.

This self-conscious leave-taking, or *exodus*, is central to what the autonomists call 'the constituent power of the multitude' (2000: 410). This is a complex concept that relies upon two components, namely, constituent power and the multitude. Reading *Empire*, it sometimes seems as though the one has simply been created for the other through rhetorical wizardry: constituent power is what the multitude wields and the multitude is what wields constituent power. At the level of revolutionary poetics—a level at which this book often operates—that does seem to be a fair reading, and a reasonable first step in interpretation. However, there is much to be gained from delving further into these concepts and the relations between them. Following the workerist line, Hardt and Negri argue that because the multitude has created Empire, it is the multitude who can bring it down. As an 'inside that searches for an outside' (185) the multitude works 'within Empire and against Empire', but not only in a negative way. These 'new' subjects also 'express, nourish, and develop positively their own constituent projects' (61). In their response to the authors who participated in a special issue of *Rethinking Marxism* devoted to critiques of *Empire,* Hardt and Negri further clarify what they see as the key components of the task of the multitude: 'resistance,

insurrection, and constituent power'. They go on to identify each
of these elements, respectively, with 'micropolitical practices of
insubordination and sabotage, collective instances of revolt, and
finally utopian and alternative projects' (Hardt and Negri 2001: 242).
Constituent power, like structural renewal, thus appears to involve
primarily the construction of alternatives; it is not about reform or
revolution. And, just as in Landauer's formulation, these alternatives
take the form of experiments which undermine Empire by draining
its energy and rendering it redundant. The tasks which might be
achieved by the constituent power of the multitude, then, are very
similar to those that have been enumerated by anarchist advocates
of structural renewal.

Although I am making this claim on the basis of theoretical
resonances, it is also interesting to note that at one point in *Empire*
Hardt and Negri suggest that the task of the multitude is to create
'a new society in the shell of the old, without establishing fixed
and stable structures of rule' (2000: 207). Here they are alluding to
the preamble to the Constitution of the Industrial Workers of the
World, from whom they say they are 'taking [their] cue' (207). The
Wobblies, of course, were the most visible and historically memorable
product of anarcho-syndicalist agitation in early twentieth-century
North America, and continue to count among their members many
self-identified anarchists. But this is not the only, or even the most
obvious, way in which autonomist theory appropriates elements of
the anarchist tradition. In an essay on the 'Virtuosity and Revolution',
Paolo Virno defines exodus as 'an engaged withdrawal, a 'founding
leave-taking' that consists in a 'mass defection from the State' (1996:
197). It is via this defection and disobedience that 'the people' are
deconstructed: 'neither 'producers' nor 'citizens', the modern virtuosi
attain at last the rank of multitude' (201). The multitude develops
a power, Virno argues, but this is a power 'that refuses to become
government' (201). Writing on his own about constituent republic,
Negri also comes out explicitly against seeking state power. 'It is
time to ask ourselves', he opines, 'whether there does not exist,
from a theoretical and practical point of view, a position that avoids
absorption within the opaque and terrible essence of the State' (219).
Negri is absolutely right in asking this question, but perhaps a little
disingenuous in pretending not only that no one has asked it before,
but that no one has come up with any answers to it. As I have shown,
anarchists since Godwin have been asking this question and providing

answers to it. Indeed, it could be said that 'the constituent power of the multitude' is nothing other than a new name for what anarchists call social revolution. 'The chief aim of Anarchism', Kropotkin wrote almost a century ago, 'is to awaken those constructive powers of the labouring masses of the people which at all great moments of history came forward to accomplish the necessary changes' (1912: 68). Add to this the anarchist mistrust of 'the people' in their mode as citizens-producers-consumers, and you end up with something very much like the autonomist multitude.

This is not to suggest, of course, that Hardt and Negri, or any other autonomist marxists, are secretly wearing circle-A T-shirts in the comfort of their own homes. 'We are not anarchists', they explicitly declare, 'but communists who have seen how much repression and destruction of humanity have been wrought by liberal and socialist big governments' (2000: 350). Yet, one cannot help but feel that this is to protest too much. Surely Hardt and Negri are aware that many anarchists, following Kropotkin, have referred to themselves as communists—marxism can by no means claim an exclusive right to the use of this term. What, really, *is* an anarchist, if not a communist who rejects the state form as a tool for achieving social change? But this definition is too simple, too easy, as are all attempts at codifying our identifications. Hardt and Negri are not anarchists, because an anarchist is someone who identifies with the traditions of anarchism, who thinks through his or her own position primarily with reference to markers drawn from this milieu rather than from some other milieu. They, and the other autonomists, I would suggest, are marxists who are quietly importing anarchist analyses and strategies, hoping to gain a certain theoretical and political purchase while avoiding censure for partaking of forbidden fruit.

Despite these borrowings, it is necessary to accept Hardt and Negri's disavowal: they are indeed not anarchists, for reasons that derive not only from their self-identification, but from their theorization of the constituent power of the multitude. While it is clear that they are aware of and positively value what I have called a politics of the act, it is not at all certain how this politics would relate to the broader project of counter-Empire. On the one hand, the multitude is theorized as a multiplicity in the Deleuzean sense, that is, as a non-identitarian formation of subjects in 'perpetual motion', sailing the 'enormous sea' of capitalist globalization in a 'perpetual nomadism' (Hardt and Negri 2000: 60–1). In this formulation, the multitude appears as 'creative constellations of powerful singularities' (61), that is, as something unknowable, untotalizable, ungraspable. It is therefore appropriate that, in another echo of post-Utopian classical anarchism, Hardt and Negri declare that '[o]nly the multitude through its practical experimentation will offer the models and determine when and how the possible becomes real' (411). At the same time, however, their language often shifts into a Hegelian mode in which the multitude appears as an entity that needs 'a center', 'a common sense and direction', a 'prince' in the Machiavellian sense (65). The philosophical answer to this conundrum of course lies in the Spinozan notion of immanence, through which the dichotomy between singularity and totality is supposed to be transcended (77–8). But the practical answer seems to lie in a rather orthodox commitment to the logic of hegemony.

This observation is based on a few scattered passages in *Empire*, but is reflective, I would claim, of a general tendency in Hardt and Negri's work. They are highly critical, for example, of Laclau and Mouffe's 'revisionist' reading of Gramsci: 'Poor Gramsci, communist and militant before all else, tortured and killed by fascism ... was given the gift of being considered the founder of a strange notion of hegemony that leaves no place for a Marxian politics' (235 n. 26). What would a properly marxist reading of hegemony look like? Hardt and Negri approvingly cite Lenin's analysis of imperialism, and give him credit for recognizing, at least implicitly, the existence of a fundamental dichotomy in modes of radical struggle: '*either world communist revolution or Empire*' (2000: 234, italics in original). In their comments on the *Rethinking Marxism* dossier, they declare themselves as being 'indebted to Slavoj Žižek for the reformulation of this question [of the ability of the multitude to make decisions] in Leninist terms' (2001: 242). It is somewhat jarring to see two autonomists reaching back behind western marxist readings of Gramsci to recover a properly leninist conception of hegemony. Yet it seems clear that the project of counter-Empire, as they conceive it, would be oriented in just this way. 'Globalization must be met with a counter-globalization', they write in *Empire*: 'Empire [must be met] with a counter-Empire' (2000: 207). Near the end of the book, they suggest that 'the actions of the multitude against Empire' already 'affirm [the] hegemony' of an 'earthly city' that is replacing the modern republic (411). This eschatological tone is maintained in a later interview, where the authors argue that 'a catholic (that is, global) project is the only alternative' (2002: 184). Finally, and perhaps most tellingly, Negri is known for this kind of approach in his own political practice, as evidenced by the comments of a fellow autonomist militant:

> Negri can be taken as an emblematic figure: every time he set foot in spaces that were opening up, in this case within the philosophical community or within the community of intellectual debate in general, he immediately tried to impose his hegemony on them or in any case force them into a hegemonic strategy. Therefore, immediately the mechanism of the party was put into play. The paradox of Autonomia was that of being born from the dissolution of the political groups only to maintain within itself the logic of the party, in other words that of the executive that had to direct, impose hegemony, address, to rein in to a common strategy and tactic everything that moved, whatever the aspect or contradiction. (Marazzi 2002)

Thus, although it may be internally differentiated and fluid, the task of the multitude—as it is envisaged by Hardt and Negri at any rate—is to counter one totalizing force with another, to struggle for hegemony in the leninist sense of this term.

Another problem with the project of the constituent power of the multitude has already been alluded to in the discussion of class-centrism above. Although at times Hardt and Negri present the multitude as a 'plane of singularities, an open set of relations, which is not homogeneous or identical with itself' (2000: 103), they also have a tendency to think of it as something singular, totalizable. '[I]f we are consigned to the non-place of Empire, can we construct a powerful non-place and realize it concretely?' (208). 'The counter-Empire must also be a new global vision, a new way of living in the world' (214). Each of these questions and statements can, and should, be rendered differently if the multitude is to be theorized as 'not a new body but a multiplicity of bodies' (2001: 243). That is: if we are consigned to the non-place of Empire, can we construct powerful non-*places* and realize *them* concretely? Or: counter-Empire must also be a *disparate but affinite set* of new global *visions*, new *ways* of living in the world. This is not a matter of mere grammar, although the language one uses in such cases is obviously important. It is a matter of the distinction between hegemonic and affinity-based forms, of the difference between a desire to build 'a coherent project of counterpower' (2001: 242) versus the desire to allow for incoherence within the ranks of those who oppose the neoliberal order, each for their own reasons.

The question being raised here is who, precisely, *is, or can be,* part of the multitude? Is the multitude perhaps identical with the 'new proletariat' (Hardt and Negri 2000: 53), understood as 'a broad category that includes all those whose labour is directly or indirectly exploited by and subjected to capitalist norms of production and reproduction' (52)? If we accept the autonomist argument that immaterial labour is becoming increasingly important, and the factory ubiquitous, then everyone, everywhere, will eventually become part of the proletariat. This seems to be the sense of the following passage:

> In the biopolitical context of Empire ... the production of capital converges ever more with the production and reproduction of social life itself; it thus becomes ever more difficult to maintain distinctions among productive, reproductive, and unproductive labour. Labour—material or immaterial, intellectual or corporeal—produces and reproduces social life, and in the process is exploited by capital. (402)

What, then, of the relationship between proletariat and multitude? Hardt and Negri don't say, but it would seem that the multitude is the proletariat made militant, the self-valorizing proletariat; to invoke an old distinction from which workerism must attempt to distance itself, it would seem that the multitude is nothing other than the new proletariat *for-itself.*

Reading the relationship between these concepts in this way helps us to understand why Hardt and Negri sometimes write as though the multitude already exists—they claim it has created Empire, for example—while in other instances they assume that it needs to be brought into being, as in the quotes above. But even on this friendly reading of their postmodern marxism, a further question is begged by the apparent ease with which the proletariat is supposed to awaken into multitude—I am referring here to the question of building solidarity across very real divisions of race, sex, sexuality, class, region, and so on. 'Cosmopolitical liberation' (2000: 64), if we can give it any meaning at all, will mean different things to different individuals and groups at different times, in different places. Some, like Hardt and Negri, will agree that state-supported proletarianization links us all; that fighting capitalism and the state form are the 'fundamental' struggles. Others will disagree, holding instead that overturning patriarchy or heteronormativity or racism is the most important task. Autonomist marxism's inability to deal adequately with these questions led, in the 1970s, to the breaking away of many of the women involved in the movement to form Lotta Feminista (Wright 2002: 134–5), and the internal feminist critique remained cogent in the 1990s (Del Re 1996). The realities of radical struggle in the postmodern condition show that *cosmopolitical liberation under a single sign is a modernist fantasy*. Total liberation does not exist, it never has existed, and it never will exist; to seek it is to give in to a Utopian urge to free the entire world once and for all, to achieve the transparent society.

This is a key insight of poststructuralist theory that Hardt and Negri refuse to take on board, and which drives their rejection of Laclau and Mouffe's deconstruction of the leninist understanding of hegemony. Although, as I have indicated, I do not agree with their turn to a liberal politics, and would push their conclusions further towards a logic of affinity, Laclau and Mouffe's work has the benefit of making it clear that we cannot simply assume that something like 'the multitude' exists, nor can we hope to bring together the multitudes under a single sign without reproducing all that is bound up with the logic of hegemony. This point has been reinforced by a number of readers of *Empire*, some of whom are otherwise quite friendly to its project. Pramod K. Mishra has pointed out that Hardt and Negri's book is 'Eurocentric in the deployment of sources, theories, knowledges, and historical events (2001: 96), and has questioned its association of the new proletariat with 'third world nomads'. Many of these subjects, he notes, 'have either become [a] miniature Bill Gates or aspire to be one' (98). This is to say that most of those who leave the 'Third World'—and certainly those who participate most closely in immaterial labour—are the elite in education, wealth and culture, and 'have no desire whatsoever to dismantle Hardt and Negri's Empire' (98). Sourayan Mookerjea makes a similar point regarding Hardt and Negri's conception of 'the global' and their consequent dismissal of 'the local': 'Is Hardt and Negri's distrust of local struggles, their inability to conceive how the defense of the local or even of national sovereignty might in specific circumstances itself be a route to 'democratic globalization' only a consequence of a surreptitious privilege given to the conditions of struggle in the United States?' (Mookerjea 2003: 2).

These critiques clearly echo those that have been brought forth in feminist contexts by women of the global South. Yet, despite its citation of some postcolonial literature, the analysis of the proletariat in *Empire* is essentializing and homogenizing; it *assumes the existence* of something that needs to be constructed, not just textually but *politically*. There quite simply *is no multitude* right now, except in the sense that there has always already been a multitude, that is, an occasionally linked, but generally disparate field of struggles with no coherence or unity. If the multitudes are ever to come together in any way, this will be the result of a long process of building solidarity and dealing with differences and structured oppressions that plague movements for radical alternatives as much as they do the political mainstream. We simply cannot wish away or have done with racism, heterosexism, classism and other forms of prejudice. Like the state form and capitalism, they are ever-present as possibilities, and therefore must be continuously acknowledged and warded off to the greatest extent possible. To put it simply: calling 'everyone' proletariat (or anything else for that matter) is to stumble blindly into a political impasse, and has the unfortunate effect of alienating precisely those with whom one might hope to build links of solidarity.

Given that they are working with a leninist conception of hegemonic social change, it should not be surprising that Hardt and Negri fail to avoid the most persistent danger of this approach. But, as I have noted, they also draw surreptitiously from anarchism, which has been working for a long time on some of the questions that seem to plague them after writing *Empire*, and which therefore might be able to offer some guidance:

> How can all this [the constituent power of the multitude] be organized? Or better, how can it adopt an organizational figure? How can we give to these movements of multitudes of bodies, which we recognized are real, a power of expression that can be shared? We still do not know how to respond to these questions (2001: 243)

At the broadest level, an anarchist response might be: you are posing yourself the wrong questions. 'All of this' is always already organized, and your 'we', whatever that might be, cannot 'give' it anything without destroying what it is. You must 'be still, and wait without hope / for hope would be hope for the wrong thing' (Eliot 1944: 28). That is, you must trust in non-unified, incoherent, non-hegemonic forces for social change, because hegemonic forces cannot produce anything that will look like change to you at all.

Here I do not want to reproduce old schisms; rather, I want to suggest that openly acknowledging the connections between anarchism and autonomist marxism can create possibilities for their mutual critique and enrichment. Harry Cleaver, for example, has pointed out the similarities between Kropotkin's theory of post-revolutionary transformation and the autonomist concept of self-valorization. 'The common element of these two approaches to the problem of transcending capitalism', he argues, 'is the search for the future in the present, the identification of already existing activities which embody new, alternative forms of social cooperation and ways of being' (Cleaver 1994). The Italian social centre movement offers many concrete examples of the kind of activity to which Cleaver is referring, as well as instances of interaction and co-operation between the two tendencies. Roberto Cimino, a militant associated with the Negri-inspired group Avanguardia Operaia, describes how the archetypal Leoncavallo social centre in Milan was founded and defended by both anarchist and autonomist activists (Cimino 1989). Other centres began to spring up in Milan during the same period. On Via Conchetta three floors of a building containing some stores were occupied, while in an adjacent area a four-floor building with some stores were similarly taken over on Via Torricelli. According to activists involved in the movement at this time, both were characterized by 'the common specificity of being promoted by a political area of certain libertarian orientation, even if it wasn't associated in any way with the "official" structures of the anarchist movement' (Aster et al. 1996: 106).[8]

It is crucial to understand that although Toni Negri is by far the best known and most influential of those involved in Italian autonomist marxism, his work, with or without Michael Hardt, hardly reflects the position of 'the movement' as a whole. As Christian Marazzi has pointed out, 'what is called the "autonomous movement" is anything but homogeneous. It is comprised of many different and sometimes opposing experiences ... Gathered here are political contributions from people who have had nothing to do with one another for years; who have chosen different political outlooks and activities' (Lotringer and Marazzi 1980: 10). Among these contributors are many who take up a line that much more definitively rejects coherence and totalization, and whose theory and practice therefore participate to a much greater extent in the logic of affinity. Paolo Virno, for example, argues that the multitude can only express itself as 'an ensemble of acting minorities, none of which, however, aspires to transform itself into a majority. ... It develops a power that refuses to become *government*' (1980: 201, italics in original). Sylvere Lotringer defines political autonomy as 'the desire to allow differences to deepen at the base without trying to synthesize them from above, to stress similar attitudes without imposing a "general line", to allow parts to co-exist side-by-side, in their singularity' (Lotringer and Marazzi 1980: 8).

Recent work by autonomists outside of Italy also reflects a growing awareness of the importance of the logic of affinity, although these writers are not always as open to acknowledging the influence of anarchist theory and practice as, say, Cleaver has been. A certain sort of coy flirtation seems to underlie John Holloway's *Change the World without Taking Power* (2002), for example. The title and circle-A on the front cover of the book lead one to believe that it is guided by an anarchist perspective, and the back cover blurb reinforces this expectation, saying that it addresses 'new types of protest movement that ground their actions on both Marxism and Anarchism'. Holloway's thesis is indeed important and worthy of notice: that we have entered a period of prolonged crisis, where instability does not provide openings for revolutionary transformation (the oppressed taking power), but rather justifies the intensification of domination and control. Under such circumstances, the silent majority don't rush out into the streets to protest, they hunker down in houses sealed with plastic to ward off the spectre of biological weapons, just as they once hid under the kitchen table in a futile attempt to protect themselves from The Big One. Because of these developments, Holloway argues, the goal of achieving state power, either through reform or revolution, has to be abandoned.

In rejecting the state form as a tool of radical social change, he does indeed follow in the footsteps of Godwin and Bakunin. It is strange to discover, therefore, that *Change the World without Taking Power* contains only two references to writers associated with the anarchist tradition, both of which are footnoted quotes from the poetry of William Blake. Anarchism as such is mentioned only twice (on p. 12 and p. 21), and in each case the references are passing. Despite its acceptance of the key insight of anarchist theory, Holloway's argument is in fact composed of extended readings of classical marxist texts, readings which reproduce such tropes as the possibility of a transparent society 'in which power relations are dissolved' (2002: 17), the state as a 'manifestation of capital's rule' (96), an 'insistence on a class analysis' (39) and a deep-seated reliance upon a knowable human nature (25), which of course 'Marx himself' got over after 1844. Unfortunately, this intervention, despite its promise interestingly to combine insights from contemporary marxism and anarchism, remains locked in a classical marxist paradigm, while at the same time stealing small peals of thunder from (post)anarchist theory and practice. Perhaps most disconcerting is the fact that Holloway flees from the task that he has set for himself, that is, to figure out how to change the world without taking power. We are historically lost, he says. 'We do not know' (215).

Nick Dyer-Witheford, a Canadian autonomist theorist, shows that 'we' in fact *do* know how to change the world without taking power, through his familiarity with a wide range of contemporary radical social movements, 'usually pejoratively and misleadingly termed "anti-globalization" movements' (2002: 2). He notes that

> While long faces on the Marxian left have been cheered by the appearance of what are now recognized—even in the mainstream press—as 'anti-capitalist demonstrators', it is equally clear that the renewed militancy is not easily ramrodded into their familiar categories. The demonstrators' diffusion of composition, diversity of perspective, decentralization of organization and, usually, determined disassociation from the disastrous historical experience of state socialism, defies the grasp of most class analysis. (3)

Dyer-Witheford reads the autonomist tradition in a much more non-hegemonic way than Holloway, or Hardt and Negri, as striving towards a 'lateral polycentric concept of anticapitalist alliances-in-diversity, connecting a plurality of agencies in a circulation of struggles' (1999: 68). He also engages much more closely and carefully with feminist and postcolonial theorists, and approves of what he calls the 'postmodern marxism' of Guattari and Negri, which has 'discard[ed] the marxist habit of nominating some agents as central to anticapitalist struggle and others as marginal' (187). At the same time, however, Dyer-Witheford's work does retain some remnants of a class-centric analysis. In a discussion of Donna Haraway's notion of the cyborg, he laments her refusal to 'nominate any central axis of conflict along which activism might be arrayed' (179). This axis is not named, but the discussion of postmodern marxism gives us a clue to its nature: it is anti-capitalist struggle, carried out now not only by workers, but by others as well. Thus, while class-centrism is avoided in Dyer-Witheford's conception of the revolutionary subject, it remains in the formulation of the *task* which this subject is supposed to take up. And, once again, anarchism and anarchists barely rate a mention; the only relevant index entry points to a glancing dismissal of what is probably anarcho-primitivism.

In sum, the theorists of autonomist marxism have made important advances in the analysis of the information economy and the society of control, through a creative application of modern marxist categories to postmodern social conditions. By taking up the anarchist critique of the party and state forms, they have pushed marxism to recognize one of its longest-standing political and theoretical impasses, and have opened this tradition up to the need for greater solidarity with other struggles that cannot be subsumed under the banner of anti-capitalism. Perhaps most importantly of all, they have shown how micropolitical struggles are not necessarily individualistic struggles; that is, they have shown how a Nietzschean (Foucauldian–Deleuzian) subject can in fact be 'social', if once the social is conceived in a way that breaks with the Hegelian tradition. At the same time, however, it would seem that the most visible autonomist theorists all maintain a basically class-centric approach. Just as Laclau and Mouffe attempt to deconstruct marxist theories of hegemony only to land in firmly liberal—that is, still hegemonic—territory, the autonomists also attempt an overcoming that ultimately falters by reverting to a leninist conception of social change that is differently, but equally, problematic. Realizing the promise of the logic of affinity requires, I have suggested, stepping out of hegemonic thinking and the revolution/reform dichotomy. This is the path being explored by postanarchism which, like autonomist marxism, also represents an attempt to rejuvenate a classical socialist tradition by passing it through the fires of poststructuralist critique.

POSTANARCHISM: A BRIDGEABLE CHASM

Before embarking on a detailed analysis of postanarchism and the logic of affinity, I want to acknowledge that many anarchists are bothered by the idea that their tradition might have anything at all in common with poststructuralist and/or postmodernist theory. John Zerzan is one prominent example of those who love to hate what they see as an 'intersection of poststructuralist philosophy and a vastly wider condition of society' (Zerzan n.d.: 1). Following a line common to many North American activists and 'critical theory'-oriented academics, Zerzan conflates the philosophical-sociological texts of Derrida, Foucault and Deleuze with the songs of Madonna, Eric Fischl's paintings and mega-malls—that is, he fails to make the necessary generic distinctions between academic theory and elite art/popular cultural products. Zerzan holds that Jacques Derrida is 'the pivotal figure of the postmodern ethos', the prophet of a 'narcissism and a cosmic "what's the difference?"' that mark 'the end of philosophy as such' (1). This is, of course, an extremely widespread opinion, as indicated in a recent interview with Derrida:

> *McKenna*: What's the most widely held misconception about you and your work?
>
> *Derrida*: That I'm a skeptical nihilist who doesn't believe in anything, who thinks nothing has meaning, and text has no meaning. That's stupid and utterly wrong, and only people who haven't read me say this. (Derrida with McKenna 2002)

As previously mentioned, it is now quite well established that deconstruction is and always has been driven by powerful ethico-political commitments. Thus it has become increasingly difficult, in the academic world at least, to avoid embarrassment while upholding the idea that poststructuralism equals postmodernism equals nihilistic relativism.[9]

The fact that Derrida, Foucault and the rest have repeatedly declared their lack of allegiance to so-called postmodernism has not, of course, done much to convince those who wish to discredit their work by attaching it unfairly to the very cultural trends they themselves deplore. Also influential among anarchist circles is Murray Bookchin's *Social Anarchism or Lifestyle Anarchism: An Unbridgeable Chasm* (1995). In this highly polemical text Bookchin laments the rise of what he calls 'lifestyle anarchism', which 'is finding its principal expression in spray-can graffiti, postmodernist nihilism, antirationalism, neoprimitivism, anti-technologism, neo-Situationist "cultural terrorism", mysticism, and a "practice" of staging Foucaultian "personal insurrections"' (19). The common link between all of these tendencies is that they appear to Bookchin as individualistic and aesthetically oriented, and are therefore 'antithetical to the development of serious organizations, a radical politics, a committed social movement, theoretical coherence, and programmatic relevance' (19). This is a list of charges which many readers will recognize, for the same fears have been raised by feminists, critical theorists and liberal philosophers. Ultimately, what worries Bookchin is the withering away of belief in a 'basic revolutionary endeavour' that would seek to liberate 'humanity as a whole' (3).

One of the prime targets of Bookchin's—and Zerzan's—wrath is Hakim Bey (aka Peter Lamborn Wilson), whom Jason Adams credits with starting all of the trouble with his 1987 essay 'Post-Anarchism Anarchy' (Adams n.d.). In this essay, Bey notes that the anarchist movement seems to be stuck between 'tragic Past & impossible Future', and therefore 'lacks a Present' (Bey 1991a: 61). Something has been exhausted, he seems to suggest, something like the romantic modernist vision of revolution of which Bookchin is so reluctant to let go, or the romantic *premodernist* vision of a society without culture that animates Zerzan's project. Although Bookchin accuses Bey of concocting a 'dreamworld' where those who enter must abandon all 'nonsense about social commitment' (Bookchin 1995: 21), Bey is in fact very aware of one of the most pressing issues facing proponents of radical social change in the 2000s:

> The anarchist 'movement' today contains virtually no Blacks, Hispanics, Native Americans or children ... even tho *in theory* such genuinely oppressed groups stand to gain the most from any anti-authoritarian revolt. Might it be that anarchism offers no concrete program whereby the truly deprived might fulfill (or at least struggle realistically to fulfill) real needs & desires? (Bey 1991a: 61)

Here Bey suggests, I think quite rightly, that the universal subject of liberation implicated in the call to liberate 'all of humanity' no longer exists, perhaps *never did exist*. But he clearly does not, as Bookchin contends, give up on social struggle as such. In fact, his vision in this 1987 essay is rather prescient: he notes that there is a great army of the disaffected roaming the streets of the G8 countries, ready perhaps to 'pick up the struggle where it was dropped by Situationism in '68 and Autonomia in the seventies and carry it to the next stage' (1991a: 62). This is precisely what we have seen happen in the 1990s—a resurgence of affinity-based radical activism that is simultaneously non-revolutionary and non-liberal. So, while Bookchin is correct that Bey urges us to abandon the quest for a 'revolution' that will achieve certain goals for 'humanity as a whole' (Bookchin 1995: 3), it is not the case that Bey advocates abandoning radical struggle as such. In fact, Bey—again quite ahead of his time—suggests that 'in the 90's we will demand effective means of *association* which depend neither on Capital nor any other form of representation'. He contends that this is the most urgent question for research and experimentation, and provides some guidance as to the directions he thinks this research *shouldn't* take: 'We reject the false trance of the Spectacular *group*—but we also reject the lonely ineffectiveness of the embittered hermit' (1991c; italics in original). Here, it would seem, is an explicit rejection of the kinds of 'individualism' offered by capitalism and the state form under the postmodern condition.

Saying what one does not want, of course, is very different from saying what one does want, and it is the latter question that is often left unanswered by those who are called 'postmodernists'. Bey, like Hardt and Negri, clearly feels that new modes of social organization are required, new ways of linking individuals with other individuals, groups with other groups. So it is a fair question to ask what he provides in the way of clues as to how this might be done. In a move that in fact sounds rather primitivist, Bey suggests that 'physical separateness can never be overcome by electronics, but only by "conviviality", by "living together" in the most literal physical sense. The physically divided are also the conquered and Controlled' (1991c: unnumbered page). On this basis, Bey seeks to resuscitate the vision of Fourier, as a 'poetics of life', in the context of a Proudhonian federalism viewed through the lens of Deleuze and Guattari's nomadology:

> Proudhonian federalism based on non-hegemonic particularities in a 'nomadological' or rhizomatic mutuality of synergistic solidarities—this is our revolutionary structure. (The very dryness of the terms itself suggests the need for an infusion of life into the theoryscape!) Post-Enlightenment ideology will experience queasiness at the notion of the revolutionary implications of a religion or way of life always already opposed to the monoculture of sameness & separation. Contemporary reaction will blanch at the idea of interpermeability, the porosity of solidarity, conviviality & presence as the complementarity & harmonious resonance of revolutionary difference. (Bey 1996: section 9)

After cutting through the poststructuralist jargon that Bey simultaneously employs and mocks, it becomes clear that he is in fact advocating *social* rather than individual change. His argument, therefore, cannot be reduced to a strain of 'lifestyle anarchism'.

Although his position is very similar to that of the autonomists, Bey is more thoroughgoing than most of them in his rejection of hegemonic forms. His theory of social change is addressed via the concept of the TAZ (Temporary Autonomous Zone), which was mentioned briefly in Chapter 1. Bey defines the TAZ as 'a certain kind of "free enclave"' (Bey 1991b: 99), a 'bit of land ruled only by freedom' (98), an insurrection that does not intend to foment a revolution or even to bring about reform. The TAZ 'does not engage directly with the State' (101); rather, it seeks to maintain its invisibility, since to become visible, to be named or 'recognized', is the beginning of the end of autonomy, the first stage of capture. Existing physically and virtually within the cracks of the societies of control, the TAZ is, as I have noted, a quintessentially postmodern tactic for social protest and the prefiguration of alternatives. At the same time, however, Bey gives the TAZ a genealogy in modern socialism, first in the experiments of the Utopian socialists, then in the revolutionary urban communes of Paris, Lyon, Munich, the free Soviets of the early days of the Russian revolution, as well as anarchist Spain. Most of these examples, of course, were revolutionary in their intent—they had clear hegemonic goals. Not so the 'pirate utopias' and 'madcap' republics (125) which Bey also holds up as examples of the TAZ. These formations differ from the others in that they display no will at all to become the state, but rather make every effort they can to stay off the maps of power, while at the same time maintaining a parasitic (piratical) relationship with the dominant apparatuses of capture and exploitation. This reveals the TAZ in its most interesting form, as an island of achieved social change, a place where the revolution *has actually happened*, if only for a few, if only for a short time.

The necessarily fleeting nature of the TAZ, however, makes one wonder whether it can do more than offer temporary respite to a small number of individuals, whether it can in fact prefigure broader and deeper social change. Bey argues—convincingly, I think—that participation in a TAZ can involve intensities that 'give shape and meaning to the entirety of a life' (100). Each moment living differently, each quantum of energy that the neoliberal societies of control do not capture and exploit, is indeed a contribution to the long-term construction of alternative subjects, spaces and relationships. However, I wonder whether the dichotomy between 'permanent revolution' and 'temporary autonomous zone' is not itself somewhat suspect. Can there not be modes of organization that are neither utterly fleeting nor totally enslaving? In a short piece on the permanent TAZ, or PAZ, Bey notes that 'not all existing autonomous zones are "temporary"', and postulates that the PAZ and the TAZ can and should feed off of one another: 'The essence of the PAZ must be the long-drawn-out intensification of the joys-and-risks of the TAZ' (1993). Here one might think of long-running intentional communities, social centres, squats, bookstores or cafés that survive while maintaining their commitment to autonomy and community. To do so, they must always be aware of the dangers of both insularity and popularity and manage, for a few years or even decades, to keep up the kind of intensity associated with the TAZ. Of course, no zone, autonomous or not, can ever aspire to total permanence; for this reason, perhaps the model that breaks us out of the temporary/permanent dichotomy is best thought of as the SPAZ, or Semi-Permanent Autonomous Zone; a form that allows the construction of non-hegemonic alternatives to the neoliberal order here and now, with an eye to surviving the dangers of capture, exploitation and division, inevitably arising from within and being imposed from without.

Despite the promise of the TAZ concept, I cannot help but share the concern that Bey's conception of social change is a little too reliant upon what seems to be an ethos of fleeting, individualistic encounters. This is undoubtedly a result of the influence of the SI which, despite its advanced positions on a number of issues, always had an air of being most amenable to young White men with no attachments to such banalities as partners, children or broader communities. Leafing through the pages of their journal, one cannot help but get the sense that those who are not willing (or able!) to spend their days drifting about the streets of Paris are doomed to act as agents of decomposition and inauthenticity, impediments to the realization of the city of the future. It is this aspect of the situationist imaginary that has led Vincent Kaufman to label the SI—in an article not entirely unsympathetic to the cause—as 'angels of purity .... invisible mortals installed between a planetary Luna Park and Never-Never Land' (1997: 66). Many of Bey's ruminations seem to come from a similar point of view: 'Whether my REMs bring verdical near-prophetic visions or mere Viennese wish-fulfillment, only kings and wild people populate my night. Monads and nomads' (1991d: 64). Installing this kind of dichotomy not only seems to run against the grain of Bey's post-revolutionary politics (is there *really* a revolutionary subject after all, and is s/he a monad/nomad?), but also leads, at times, to uncritical celebration of qualities that are assumed to be associated with nomads.

> La décadence, Nietzsche to the contrary notwithstanding, plays as deep a role in Ontological Anarchy as health—we take what we want of each. Decadent aesthetes do not wage stupid wars nor submerge their consciousness in microcephalic greed and resentment. They seek adventure in artistic innovation & non-ordinary sexuality rather than in the misery of others. (1991d: 44)

Decadent aesthetes do not wage stupid wars? Has Bey not heard of fascism? They don't fall into greed and resentment? It seems that more than a few monads (absolute monarchs) have gone precisely this way; and isn't *everyone* who gets caught up in capitalist production and consumption ultimately seeking a life of 'pure' pleasure, beyond commitment and care? Doesn't *everyone* want to be a monad nomad?

Although Bey's style particularly lends itself to this kind of rhetoric—and more (less?) power to him, he is a good poet and generates interesting concepts—the nomadic subject is to be found wandering through many other texts that operate at the intersection(s) of anarchism and poststructuralism. Rolando Perez, in an early effort at working through links between schizoanalysis and what he calls an(archy), has argued that both of these practices aim at 'the replacement of poor defenseless, guilt-ridden puppets in internal strait-jackets, with free, non-Oedipalized, uncoded individuals' (Perez 1990: 28). Once again, this sounds wonderful, but given that 'the individual' is a discursive construct associated historically with western bourgeois liberalism and its conception of the family, how could an 'individual' exist in a 'non-Oedipalized, uncoded' form? Indeed, how could any subject do as Perez suggests, and 'destroy his or her own form of expression immediately, so as to make repetition and incorporation impossible?' (57). Although Perez relies heavily upon the work of Deleuze and Guattari, he seems to forget that repetition (in the Deleuzean sense) is unavoidable, and incorporation is always a danger to be warded off. When thinking about the possibilities of nomadic subjectivity, the final sentence of *A Thousand Plateaus* should always be kept in mind: 'Never believe that a smooth space will suffice to save us' (Deleuze and Guattari 1987: 500).

Lewis Call's 'anarchy of the subject' (2002: 22) is similar to Bey's and Perez's conceptions, but with a crucial difference: he is aware of a danger of 'pure' nomadism that they do not seem to recognize. 'If all essence, all fixed being, all laws of states and subjects are to be swept away in the torrent of becoming, can we be sure that this torrent will not carry us into some dark quagmire? Can we avoid, for example, the danger of becoming-fascist?' (52). This is to raise the *question* of becoming, rather than to simply *assume* that becoming is somehow innately superior to 'mere' being. Once again, Braidotti's work on nomadic feminism is helpful. She notes that Deleuze and

Guattari do not suggest that "homelessness' and 'rootlessness' are the new universal metaphors of our times. This level of generalization is not of much help' (2002: 84). It is not of much help because it fails to make a distinction between becoming-major (seeking hegemony) and becoming-minor (taking a non-hegemonic line). Not only this, but becoming-man, as Braidotti notes, is a very different thing from becoming-woman/child/animal, or any of the endless series of positions that are aligned around man as its supposedly inferior others. Thus, as she argues, 'the politics of location is crucial' to any nomadic becoming (84). Where one begins, where one is going, how one plans to get there, are unavoidable *political* questions if one wants to take a line of flight from the neoliberal (or any other) order. So, while we all know that it is possible to maintain an affinity for a couple of days merely on the basis of a common desire to throw off momentarily the shackles of the working week, figuring out a way *not to have to go back to work* is a much more difficult task. It is, in fact, a task that requires sharing certain values, strategies and tactics. Call does not take the discussion very far; he simply suggests that 'microfascism should be understood as the limit which defines becoming, grants it a definite (albeit fluid and flexible) shape, and prevents it from dissipating into a politically meaningless gasp of chaos' (2002: 52–3).

Why and how microfascism might play this salutary role is not explained, leaving the impression that more work needs to be done on the ethico-political commitments that guide postanarchist theory and practice. For clues as to how this work might proceed, I want to turn to Todd May's *The Political Philosophy of Poststructuralist Anarchism* (1994). In this early and influential book May develops a number of lines of analysis that have become central to postanarchist critique: he rejects both 'top-down' (classical marxist) *and* 'bottom-up' (classical anarchist) theories of revolutionary transformation, in favour of a network analysis; he is critical of anarchism's assumption that 'the human essence is a good essence, which relations of power suppress or deny' (62); he notes the links between anarchism and *autonomia*, as well as the problem with the class-centric nature of much autonomist theory. Unlike the other postanarchist writers discussed so far, though, May also takes on the problem of ethics at the heart of his inquiry, presenting what will be familiar as a Habermasian critique of the 'performative contradiction' that is supposedly inherent poststructuralist ethics:

> The mistake, made by Deleuze and Foucault in avoiding ethical principles altogether and by Lyotard in trying to avoid universalizing them, is that their avoidance is itself an ethically motivated one. (130–1)

The problem with this reading is that it fails to respect an important distinction between ethics and morality, a distinction May himself recognizes at other points in his argument. In a discussion of the concept of 'experimentation' in Deleuze, for example, he notes, in a way similar to Braidotti, that 'the task of becoming-minor is precisely that; it is not a task of making the minor dominant' (115). In the final chapter on 'Questions of Ethics', however, this distinction is blurred. Regarding Foucault's ethic of the care of the self, May parenthetically warns the reader that 'Foucault here is using the term "ethics" to denote a practice of self-formation, while our use of the term is more traditional, referring to binding principles of conduct' (123). But this issue cannot be kept in parenthesis, for the particularity of Foucault's ethical position relies precisely upon a break with the 'traditional' usage. As Deleuze notes, 'establishing ways of existing or styles of life isn't just an aesthetic matter, it's what Foucault called ethics, as opposed to morality. The difference is that morality presents us with a set of constraining rules of a special sort, ones that judge actions and intentions by considering them in relation to transcendent values (this is good, that's bad ...); ethics is a set of optional rules that assess what we do, what we say, in relation to the ways of existing involved. We say this, do that: what way of existing does it involve?' (Deleuze 1995: 100).

The suppressed privilege of morality over ethics comes into the open during a discussion of *Just Gaming*. Here, May notes, Lyotard argues that 'any discourse meant to account for prescriptions transforms them into conclusions of reasonings, into propositions derived from other propositions' (cited in May 1994: 129). Rather than accounting for prescriptions, Lyotard suggests, the point is to respect the particularity of language games and minimize the creation of differends. Forgetting the earlier distinction between becoming-minor and becoming-major, May now finds this position to be 'internally incoherent', since 'almost all ethical principles involve caveats' (131). With the appearance of the familiar argument of 'performative contradiction', the *dénouement* is not far off. May writes: 'The picture of ethical discourse we want to develop here is one that takes it as a practice of making, endorsing, and discussing claims that involve values and practical judgments, the commitment to which is, or at least ought to be, given the by weight of the best reasons on behalf of those values or practical judgments' (141). Thus a text which contains some important work on points of contact between anarchism and poststructuralist theory diverges into a consideration of analytic, rational-universalist formulae such as whether 'one ought to perform action X under circumstances C', or indeed how we shall know whether 'Circumstances C obtain' (149).[10] By the final curtain, Lyotard, Deleuze and Foucault have been hopelessly crushed under the weight of a descending Habermasian machine.

The question of ethics is taken up in a more immanently poststructuralist register by Saul Newman, whose reading is not so strongly guided by the presumptions of Critical Theory. Newman starts from a position similar to May's, suggesting that poststructuralism 'offers little possibility of a coherent theory of political action' (2001: 159), because it does not provide a 'way of determining what sort of political action is defensible and what is not' (160). The use of the passive voice here is telling: Newman seeks an *argument*, an appeal to a rational-universal subject. But a few pages later, he manages to escape the trap of morality with the help of Derridean deconstruction and Laclau's concept of the empty signifier. I won't go into the technical details of this argument, as I am primarily interested here in theory rather than metatheory;[11] it will suffice to note that Newman does acknowledge the distinction between morality and ethics outlined by Deleuze:

> Poststructuralism rejected morality because it was an absolutist discourse intolerant of difference: this is the point at which *morality becomes unethical*. Ethics, for Derrida, must remain open to difference, to the other. (2001: 167; italics added)

While avoiding one pitfall, however, Newman falls into another, which is hinted at in his suggestion that poststructuralism and anarchism share 'a commitment to respect and recognize autonomy and difference' (170). The appearance of key terms from the discourse of liberal multiculturalism ('respect and recognize') alongside a key term from anarchism and autonomist marxism ('autonomy') should give us pause. Is Newman trying to turn postanarchism into nothing more than another flavour of hegemonic politics? It seems so. On the following page he notes that 'we seem to be surrounded today by a multitude of new identities ... a new proliferation of *particularistic demands*' (171; italics added). One begins to feel apprehensive about the arrival of another familiar vehicle—Laclau and Mouffe *ex machina*? The penultimate page spells it out: 'This book has attempted to make radical anti-authoritarian thought more "democratic"' (174). Why does the word democratic appear in quotes? Another book could be written about that, but it should be clear by now that 'democratic' here means something like 'properly oriented to a postmarxist liberal politics of demand that does not challenge the state form as such'. This is *not* the kind of postanarchist politics I would advocate, any more than Habermasian discourse ethics is the kind of postanarchist ethics that we need.

Here, of course, I beg the question of what kind of politics, ethics— and why not, let's add subjectivities—*are* desirable. For this I would recall the discussion of poststructuralism with which this chapter began, and return to Foucault. Like Laclau and Mouffe, Foucault adheres to the central insight of what we might call Nietzschean sociology: relations of power are inherent to human societies, and it is at best futile, at worst an invitation to totalitarianism, to wish them away. But Foucault goes further than deconstructive/Lacanian political theory, by providing an analytics of power and an ethic of care for the self that allow us to differentiate between the various modalities of power relations. 'The problem is not one of trying to dissolve [relations of power] in the utopia of a perfectly transparent communication', Foucault writes, 'but to give one's self the rules of law ... the *ethos*, the practice of self, which would allow these games of power to be played with a *minimum of domination*' (1987: 129; italics added). This formulation begs several questions: first, how can a relation of domination be distinguished from other relations of power? And second, just what are these rules that one can give oneself in the name of minimizing relations of domination?

Foucault's analytics of power provides an answer to the first question, by way of the play he sets up between relations of power as 'strategic games between liberties' and 'states of domination' (130). In the first case, we have what we might call 'live' relations of power, in that most of the players, most of the time, have some ability to alter the situations in which they find themselves; in the second, the flow of power has 'congealed' or been 'blocked', preventing 'reversibility of movement' for some of the players most of the time (114). In these situations, which are brought about through the use of specific 'techniques of government' (130), individuals are confronted by what we might call 'dead' power. But this is not the end of the game; at this precise point a third type of power relation must be considered, that of 'struggle' or 'resistance', which Foucault argues is 'necessarily' found wherever there is domination (123). Now, while his analyses portray all three of these modalities as mutually interpenetrating, Foucault does not see them as axiologically equivalent. For him, the role of the intellectual is to participate, along with others, in a 'struggle against the forms of power that transform him [or her] into its object and instrument ... a struggle aimed at revealing and undermining power where it is most invisible and insidious' (Foucault and Deleuze 1976: 75).

These 'local and regional' (75) practices of resistance are one way in which individuals and groups can work against relations of domination. But they are not the only way. Within the ethic of care for the self a further modality can be found, that of exerting *control over oneself*, so that one does not give in to an urge to exercise 'tyrannical power' over others (Foucault 1987: 119). Foucault achieves a curious, but crucial, inversion of the Christian paradigm: one gives oneself an ethic so as not to succumb to the *temptation* of morality. It is important to note that, with this recourse to ancient ethics, Foucault is not calling for the return of something lost in the past; rather, he suggests that the ethic of care for the self could be of use in helping to produce 'something new' in philosophy and politics, by offering an alternative to a modern-Christian ethic of self-sacrifice in the name of care for others (115–16). This something new, what could it be like? There are clear affinities between Foucault's commitment to resistance to domination and the deconstructive themes of openness, the *à venir*, and contingency. But again, the Habermasian sceptic is sure to note, we have the characteristic orientation only to what *cannot* be said or done. The question remains: does poststructuralism have anything 'positive' to say about the possibilities for social action and transformation?

The works of Deleuze and Guattari are of help here, as they offer the boldest—and hence the most 'dangerous'—forays into *constructive* social criticism to be found among the writers commonly considered as poststructuralist (MacKenzie 1997). In a certain mood, ethical injunctions fly from Deleuze and Guattari's texts as though seeking respite from the heavy burden of perspectivist rigour. 'Always follow the rhizome by rupture', we are told; 'lengthen, prolong, and relay the line of flight' (Deleuze and Guattari 1987: 11). '[I]ncrease your territory by deterritorialization' (11). 'We should stop believing in trees, roots, and radicles' (15). 'Don't bring out the General in you! ... Make maps, not photos or drawings. Be the Pink Panther' (25). In another mood, Deleuze and Guattari issue somber warnings about the kind of ecstasy into which they themselves are prone to fall. They write: 'If it is a question of showing that rhizomes also have their own, even more rigid, despotism and hierarchy, then fine and good: for there is no ... ontological dualism .... between good and bad, no blend or American synthesis' (20).

Reason fails in Deleuze and Guattari's texts, though it does not fail miserably. It fails *joyfully* and *playfully*, in the same way that the Pink Panther paints the world his own colour. Both ecstasy (passion, Bey, nomad, SI) and caution (reason, May, citizen, Habermas) are set loose, neither subordinated to the other. But this does not mean that 'anything goes': Deleuze and Guattari never exhort their readers to 'Become a man of the State', 'Do Royal Science' or 'Get Oedipal!' A crucial aspect of Deleuze's philosophical method is his commitment to a style of criticism that proceeds by the creation of alternatives: alternative readings, concepts, planes of immanence. Deleuze and Guattari's *social* criticism, as MacKenzie has argued, similarly 'involves the creation of new concepts of society', or at least 'the creation of new concepts pertaining to social relations' (1997: 13; 17 n. 41). Thus, although they are careful to point out that there are no general recipes or globalizing concepts to which one can turn with certainty in every case, Deleuze and Guattari's analyses of contemporary western societies tend to identify certain excesses and provide suggestions as to how these excesses might be fought, repaired, or partially escaped.

But what is to be fought, and why? While Deleuze and Guattari are careful to avoid ontological dualisms that would precede and motivate ethico-political choice and social analysis, they utilize a network of contingent dualisms that enable their critique of particular systems of power relations. I will focus here on only one linked subset of concepts that resonates with, and adds further complexity to, Foucault's work on the analytics of power. Foucault, as I have noted, marks a distinction at the level of system—that is, at the level of flux, flow, process—between relatively open and relatively blocked relations of power. Deleuze and Guattari provide further help in seeing our way past the hegemony of hegemony with their insights into relations between the state form and the war machine. States tend to perpetuate already instantiated (arborescent) forms, while war machines tend to destroy old forms and instantiate new ones through rhizomatic connections. Thus, for Deleuze and Guattari, 'revolutionary organization must be that of the war machine' (Guattari 1995: 66);

indeed, they see their own writing as an operation that 'weds a war machine and lines of flight, abandoning ... the State apparatus' (Deleuze and Guattari 1987: 24). Yet, just as arborescent forms can grow rhizomatic appendages, states can—and must—incorporate war machines, tame them and put them to use in 'an institutionalized army', make them part of the 'general police' function (Deleuze and Parnet 1983: 103). This is the 'special danger' of the war machine, seldom noted by those who have too easily appropriated the rhizome concept: if it does not succeed in warding off the development of a state form, it must pass into the service of the state or destroy *itself* (104). Inhabitants of the SPAZ, beware!

It is here, in the form of ecstatic injunctions accompanied by somber warnings, that Deleuze and Guattari, like Foucault, present not only a 'negative' call to resistance, but also a consistent and 'positive' ethico-political stance. At times, they take us even further than this, advocating what Keith Ansell Pearson has called 'novel images of positive social relations' (1998: 410). Thus Deleuze: 'We have no need to totalize that which is invariably totalized on the side of [dead] power; if we were to move in this direction, it would mean restoring the representative forms of centralism and a hierarchical structure. We must set up lateral affiliations and an entire system of networks and popular bases' (Foucault and Deleuze 1976: 78). This system of networks and popular bases, organized along rhizomatic lines and actively warding off the development of arborescent structures, could only be populated by subjects who neither ask for gifts from the state (as in the postmarxist, liberal-democratic new social movements) nor seek state power themselves (as in classical marxism). They would have to be aware of the dangers of creating a non-statist, but still hegemonic totality (as in leninist versions of autonomist marxism), and of seeking only fleeting Rave-like experiences of individual 'liberation' (as in the TAZ). These molecular movements would need to resist the will to domination in Foucault's sense, that is, they would need to take up ethico-political positions while refusing to try to *coercively generalize* these positions by making foundational claims. Rather, they must be content to find allies where and how they find them, for as long as they find them.

CITIZEN, NOMAD, SMITH

Shifting our identifications is never easy, and this must surely be one of the most profound reasons that most people, most of the time, do not desire radical social change. They/we would prefer to remain as 'citizens', that is, as Oedipalized subjects making demands of those in authority: daddy-mommy-me easily morphs into state-corporation-me, First World-Third World-me, and so on. The nomadic subjects theorized by autonomist marxism, postanarchism and postmodern feminism are compelling because they represent attempts to abandon parental protection finally and absolutely. These theorizations draw heavily from narratives associated with the state form, in which nomads appear as the negation of all of the qualities attributed to the citizen, as Barbarians who 'sow not, nor have any tillage ... [are] without habitation, having no dwellings but caves and hollow trees' (D'Avity, cited in Hodgen 1964: 201). In an archetypal nightmare of European civilization, the nomadic war machine gallops in off the steppes, sweeping away everything that matters: fields, walls, houses, castles. That the East has fared no better is suggested by that monument to state insecurity, the Great Wall of China.

What is it about nomads that makes them so frightening to sedentaries? Deleuze and Guattari provide us with a whole series of dichotomies between these two modes of subjectivity, but the most important for my purposes here is the difference between the kinds of space that they occupy. Citizens are at home in the striated space of the state form, while nomads occupy the smooth spaces of non-state relationships. As with all of the concepts deployed by Deleuze and Guattari, no stable definitions or oppositions can be sustained with respect to these two spaces—yet it is clear that 'they are not of the same nature' (1987: 474). On the plane of technology, striated space is associated with fabric, while felt is considered smooth; using a maritime model, the ocean prior to the invention of latitude and longitude was extremely smooth, that is, one navigated according to 'wind and noise, the colours and sounds of the seas' (479); finally, we can think literally in terms of territory, and distinguish between the smooth space of the North American prairie prior to European colonization and the striated spaces of its division into a firm, grid-like structure, fenced off against errant flows of flora, fauna, and indigenous peoples.

But, I have suggested above, it is impossible to be fully deterritorialized and still remain a subject—this is the limit of psychosis, a form of 'absolute freedom' that obliterates the subject in the moment that it attains its goal. The positions of the citizen and the nomad are, and must be, deeply interrelated, to the point of reversibility. The inside and outside of any social space are *interdependent*, each potentially giving rise to the other, each warding off the other, in an ongoing play of relations of co-operative and competitive power. Just as the state and the war machine are in perpetual interaction, so the citizen and the nomad share a space of contested de- and re-territorialization, each attempting to instantiate and perpetuate the conditions of possibility of its own style of life. For example, the USSR appeared to the USA as a horde of communist barbarians poised at the gates of civilized capitalism, while at the same time the Soviets feared the immense deterritorializing power of capital. The same thing is happening now between the USA and Islam, or really, if one looks closely at the arrangement of friends and enemies, with the USA and all that is not yet sufficiently USA. The beauty of this formulation is that friends can be turned into enemies and vice versa as need be, for example in the demonization of Canada and the glorification of Poland as a result of the exigiencies of the US/UK invasion of Iraq in 2003. And, as the world is progressively Americanized, the criteria for acceptance as properly pro-American become ever tighter. We can already see the coming of a time when most Americans themselves will be excluded, but this will hardly matter, for by then America and the Rest of the World will be indistinguishable. There will be no more outside, as Hardt and Negri argue—or will there?

The question of the total conquest of the outside leads us to consider a third mode of subjectivity analysed by Deleuze and Guattari, one that has not caught on in the same way as the nomad. I am thinking here of the *smith*, who exists in a complex relation to both the citizen and the nomad, in one aspect as their complement: 'There are no nomadic or sedentary smiths', Deleuze and Guattari write. 'The smith is ambulant, itinerant: his space is *neither* the striated space of the sedentary, *nor* the smooth space of the nomad' (1987: 413; italics added). In another aspect, the smith takes up a contradictory position: '[I]t is by virtue of his itineracy, by virtue of his inventing a holey space, that he necessarily communicates with the sedentaries *and* with the nomads (and with others besides ...). He is in himself a double: a hybrid, an alloy, a twin formation' (415; italics in original). Where the practice of the citizen is oriented to 'staying on the road', as it were, and that of the nomad to destroying all roads, the smith is guided by an alchemical, metallurgical will to the 'involuntary invention' (403) of new strategies and tactics. Rather than attempting to *dominate* by imposing all-encompassing norms, the smith seeks to *innovate* by tracking and exploiting opportunities in and around existing structures. The figures of the hacker, the monkeywrencher and the invisible hero of Italo Calvino's *If on a Winter's Night a Traveler* ... all come to mind. Even Royal Science—military, political, sociological, bureaucratic—has often made its 'discoveries' through this method, though it has seen fit to provide itself with a sedentary myth of control and purposeful advance. The key point I want to make here, though, is that the activities of smiths show us that no matter how totalizing a system might be, it will never achieve its ambition of totality—it is impossible to create a system with no outside, even a system that appears to cover an entire planet. For there will always be holes, even when there are no longer any margins. And out of these holes will spring all manner of subjects.

To avoid the lure of totalizing differentiations, we must note that the smith is *not* a product of a 'good' essence, or any 'essence' at all. The logic of the smith is rather a 'pure possibility, a mutation', as 'the borrowings between warfare and military apparatus, work and free action, always run in both directions, for a struggle that is all the more varied' (403). Through his or her participation in a mass-technocratic, risk-based society, the postmodern citizen becomes adept at estimating magnitudes of known quantities and undergoes what might be called a *rational becoming-subject*. Just as sexualized subjects rely upon identity-producing performances (Butler 1990), politicized subjects repeat general, timeless and deterministic procedures as laid down by a script which is rarely exposed to modification. Among these scripts I would include those which guide attempts at reform and revolution. In contrast to the rational becoming-subject of the citizen, the smith experiences an *arational becoming-object*, through jarring encounters with the social-political real—with modes of social existence which cannot, must not, signify. By this I mean to refer to those practices—anarchist, indigenous, queer, feminist—which have been submerged for several centuries under a complex and ever-changing hegemony of (neo)liberal and (post)marxist forms, practices which are now re-emerging as the limits of reform and revolution become ever more apparent.

But, it will be noted, the catastrophic failures of twentieth-century social experiments—totalitarianism under actually existing socialism, crime, poverty, racism and heterosexism in the Free World—are precisely what have motivated both the new social movements and postmarxist theory. What sense could there be in arguing for more and varied experiments leading to more and varied failures? To answer this question a further distinction must be developed, between the paranoid subject of the *mass* and the schizo subject of the *pack*. 'Among the characteristics of a mass ... we should note large quantity, divisibility and equality of the members, concentration, sociability of the aggregate as a whole .... Among the characteristics of a pack are small or restricted numbers, dispersion ...' (Deleuze and Guattari 1987: 33). Clearly, an experiment carried out as part of a mass movement is much more dangerous than the same experiment undertaken by one or more packs. In arguing for further attempts and failures, I am advocating neither mass revolution nor mass reform, but small-scale pack wanderings through the infinite social spaces that are left unexplored by these predominant models. In this sense, the smith could be thought of as a kind of *communitarian nomad*, as an intensity that burns only, or at least burns best and longest, when alongside others.

In closing this chapter, I want to make it clear that I am *not* denying the utility of citizenship for achieving certain sorts of change within ostensibly 'liberal' societies. Nor am I willing to suggest that armed revolutionary struggle is inappropriate in situations where not even the rudiments of a 'liberal' political order exist. Nor, even, am I suggesting that those individuals who *can* 'free' themselves for a few hours or days should not seek to do so. Only the Oedipal subject can work within Oedipal societies, only the revolutionary subject can overthrow blatantly totalitarian regimes, and only the nomad can escape them both for a time. None of these spaces, and the subjects that inhabit them, will be disappearing anytime soon; indeed, if Deleuze and Guattari are correct in seeing them as variants of forms that have always been with us, they can never be expected to disappear entirely. What I want to argue is that continuing with an *exclusive* focus on hegemonic change via the state form, or on escaping it entirely, prevents us from imagining and implementing modes of social organization that are not only possible and desirable, but are becoming ever more necessary as Empire consolidates its hold on our bodies, minds, lands ... on our very ability to produce ourselves and the contexts in which we encounter others. These modes can only be explored by *relatively* de-Oedipalized subjects who are able to act, without necessarily having state sanction or support, in the gaps between, and on the margins of, the institutions of sedentary society; subjects who do not love the state form, but can co-exist with it if they must, as they seek to render it increasingly redundant; subjects who seek to avoid microfascisms, who practise an ethic of care of the self, but who are also open to sharing values, resources and spaces with others, to building communities of resistance and reconstruction that are wider and more open, yet remain non-integrative in their relation to others. The movements, groups and tactics which I discussed at the start of this book are all examples of this kind of subjectivity, of these kinds of spaces, which rely upon an amoral, postmodern ethics of shared commitments based on affinities rather than duties based on hegemonic imperatives. These commitments are necessarily always shifting, but also always present, as no community can be sustained without them.

6

Ethics, Affinity and the Coming Communities

Twenty-one years ago we struggled with the recognition of difference within the context of commonality. Today we grapple with the recognition of commonality within the context of difference.

(Anzaldúa 2002a: 2)

WHAT ARE THE COMING COMMUNITIES?

A community composed by affinity-based relationships is not a Hegelian *Sittlichkeit,* nor is it a brand of liberal/postmarxist pluralism. It is not even what the more hegemonically oriented autonomists think of as 'the multitude'. All of these conceptions gloss over too many real differences and struggles that are encountered by those trying to come together against neoliberalism, while inhabiting disparate regions, positions in political-economic structures and racial/cultural/sexual identifications. Even the logic of affinity as it emerged within classical anarchism, and as it has been taken up by postanarchists, still lacks something crucial—adequate attention to axes of oppression based on practices of division rather than of capture or exploitation. Certainly, some autonomist and postanarchist writers have made an effort to address the work of theorists and activists who are not necessarily identified with their own traditions. But to begin to realize what Giorgio Agamben (1993) has called the 'coming community', it will be necessary to engage more directly, more deeply, more fully, with subjects and struggles that have been marginalized for too long by the 'big three' political discourses. In this chapter I will discuss how Agamben theorizes the coming community, and suggest that his conception needs to be modified in order to withstand an obvious critique on the very ground he claims as his own: we must speak of the coming *communities,* in the plural, if we want to go as far as possible in warding off what may appear as a hegemonic moment. As postmodern and anti-racist feminists have argued, a multidimensional, interlocking analysis of oppression is crucial to an adequate understanding—and undermining—of the neoliberal project. But this alone is not enough. In addition to theorizing about a non-identical identity that is assumed to already exist, at least *in potentia*, it is necessary to find more ways to link actually existing groups through a shared commitment to groundless solidarity driven by infinite responsibility. To the extent that this commitment drives concrete action, to the extent that it brings about changes in daily practices, obstacles based on traditional divisions can be overcome. This is, of course, an endless process, but is essential to creating and maintaining the affinity-based relationships that compose the coming communities.

Just as writers like Foucault and Deleuze are commonly supposed to spurn ethical and political commitments, it is also often assumed that they are opposed to any kind of community. Many of those who take up this position are hegemonic thinkers who are clearly frightened by the implications of non-hierarchical relationships for the liberal/ critical-theoretical status quo (Taylor 1984; Fraser 1981: 273). They see anything other than hegemonic (Hegelian) community as no community at all. Others offer versions of Spivak's critique, associating those who do not orient to a singular community within which representations might be made with 'a violently representational colonial ethnography' (Jardine 1985: 217; Hartsock 1990); that is, critics of the recognition paradigm are said to be upholding White male privilege at the same time as they shirk the responsibilities that go along with it. Others, like Peter Hallward, have pushed the concept of what I have called the 'pure' nomadic subject to its theoretical limits, beyond even the positions of writers like Hardt and Negri, Braidotti, Perez or Newman. Hallward suggests that 'Deleuze works very literally toward a world *without* others altogether; that is, he denies the philosophical reality of all relations—with and between others' (1997: 530).

These readings are, I would argue, rather tendential—that is, they emphasize one line of highly abstract and philosophical argumentation while ignoring concrete, political implications and direct statements that provide an essential counter-reading. Deleuze, for example, has noted that:

> [s]ubjectification wasn't for Foucault a theoretical return to the subject but a practical search for another way of life, a new style. That's not something you do in your head .... Where and how are new subjectivities being produced? What can we look for in present-day communities? (1995: 106, 115)

Foucault's own comments resonate with Deleuze's understanding of his approach, and also reinforce his commitment to non-hegemonic modes of constructing communal identifications:

> [T]he problem is ... to decide if it is actually suitable to place oneself within a 'we' in order to assert the principles one recognizes and the values one accepts; or if it is not, rather necessary to make the future formation of a 'we' possible, by elaborating the question. Because it seems to me that the 'we' must not be previous to the question; it can only be the result—and the necessarily temporary result—of the question as it is posed in the new terms in which one formulates it. (Foucault 1997b: 114–15)

Examples such as these could be multiplied, but I will assume this is not necessary, given the work done in the previous chapter to elaborate upon the distinction between ethics and morality in poststructuralist theory. Just as the rejection of coercive morality need not necessarily lead to passive nihilistic relativism, so the rejection of Hegelian community need not necessarily lead to an anti-social individualism. In poststructuralist theory, it leads to something quite different that can be approached via the concept of *singularity*.

This concept provides relief from a number of dichotomies that have long plagued western social and political thought. In the context of political organization, it breaks down hard-and-fast distinctions between the individual and the community, the particular and the universal. As Agamben notes, singularity is 'freed from the false dilemma that obliges knowledge to choose between the ineffability of the individual and the intelligibility of the universal' (1993: 1). What Agamben calls 'whatever being' is that aspect or moment of being that is relatively free from dependence upon identification and subjectification, from the poles of the mass, the many, the well-disciplined, the people. What is this sort of being, apart from its abstract expression? Agamben says it is *the example*. 'Neither particular nor universal, the example is a singular object that presents itself as such, that *shows* its singularity' (9). Thus, rather than proceeding from the universal to the particular, or the particular to the universal, Agamben argues that both the universal and the particular emerge out of whateverness.

> It is from the hundred idiosyncrasies that characterize my way of writing the letter p or of pronouncing its phoneme that its common form is engendered.

Common and proper; genus and individual are only the two slopes dropping down from either side of the watershed of whatever (20)

Agamben's analysis wonderfully undercuts the basis of 'socialization' in the functionalist sense, by denying any primacy to 'the normal', or to any patterned distribution of supposed 'human traits' at all. 'Taking place, the communication of singularities in the attribute of extension, does not unite them in essence, but scatters them in existence' (19). In everyday terms, whatever being is what causes/ allows subjects to resist the systematic imperatives, both overt and covert, that attempt to structure their lives; it is what breaks us out of the societies of discipline and control, and urges us towards creating our own autonomous spaces. Whatever being also compels us to act ethically in the poststructuralist sense; it compels us to *make choices* under circumstances where it is impossible to relieve ourselves of responsibility by an appeal to moral necessity.

Whatever being in the coming community constitutes, of course, a politics, a set of interventions in linked fields of power/knowledge. Agamben argues, in keeping with the line I have been advancing in this book, that 'the novelty of the coming politics is that it will no longer be a struggle for the conquest or control of the state, but a struggle between the state and the non-state (humanity), an insurmountable disjunction between whatever singularity and the State organization' (85). Here Agamben takes up the anarchist distinction between social and political revolution and, like the autonomist marxists (with whom he has close connections) comes down on the side of the former. He is also aware of the problems associated with liberal multiculturalism, arguing that whatever singularities must not form a 'societas'; if they do, they become vulnerable to recognition and integration. 'What the state cannot tolerate in any way', he remarks, 'is that the singularities form a community without affirming an identity, that humans co-belong without any representable condition of belonging' (85).

This observation, of course, begs the question of how the coming community might be expected to arrive. On this question Agamben takes up a characteristically marxist position, arguing that capitalism, and especially its advertising technologies, through their commodi-fication of the 'image of the body' (50), have given us a glimpse of whatever-being as an infinitely malleable consumer-body, ready to take in and or put on *anything*. Given that this form of subjectivity already exists, Agamben suggests that the pertinent task is to wrest it

from the circuits of capital: 'To appropriate the historic transformations of human nature that capitalism wants to limit to the spectacle,
to link together image and body in a space where they can no longer
be separated, and thus to forge the whatever body, whose physis is
resemblance—this is the good that humanity must learn how to wrest
from commodities in their decline' (50). To get over capitalism, then,
Agamben suggests we must push it to its limits, force its contradictions. The same goes for the state form. Agamben argues that all of
the 'kingdoms of the earth' are setting a course for the 'democratic-
spectacular regime that constitutes the completion of the state-form'
(83), for what Hardt and Negri call Empire. He argues further—again
in a mode similar to that of Hardt and Negri—that 'only those who
succeed in carrying it to completion ... will be the first citizens of a
community with neither presuppositions nor a state' (83).

Although Agamben's concept of the coming community is clearly
important for understanding how affinity-based social forms might
be organized, it poses a number of problems that must be addressed.
First, the very language that Agamben uses is problematic. By consistently referring to 'the coming community' using the singular
form, he seems to imply that despite all of the disparities and lacks of
which it is composed, this new mode of association will be totalized
at some level. This reading is supported not only by the language
used, but in the assumption that the coming community represents
the 'completion' of spectacular capitalism and the state form. This
kind of argument resonates strongly with modern ideas about the
Revolution leading to the transparent society and the end of history,
ideas that, as I have shown, have been mostly abandoned by contemporary theory and activism. If the coming community happens
to arrive—or, more precisely, if it begins to arrive more quickly, fully
and deeply—history will continue, state and corporate forms will still
exist, and we will remain divided in many ways—nothing will have
been 'completed', since nothing *can be* completed. Only if we want
to lurch from one hegemonic system to another will we long for a
mythical world in which 'the Shekinah will have stopped sucking
the evil milk of its own separation' (83). Rather than longing for
total communion, we must understand communities as multiplicities that cannot be totalized, as *n*-dimensional networks of networks
that spread out infinitely and are infinitely interconnected. We must
always speak of the coming communities in the plural form, and
forget about the Spinozan trick that allows us to think that *our* way

of sewing everyone up into the same bundle somehow avoids the problem of suffocation.

It is on this basis that I would disagree with Agamben regarding who is likely to build the coming communities and, more importantly, who is already building them. I don't think it's those who are most wrapped up in consumer capitalism and the state form. Rather, I think that the coming communities are more likely to be found in those crucibles of human sociability and creativity out of which the radically new emerges: racialized and ethnicized identities, queer and youth subcultures, anarchists, feminists, hippies, indigenous peoples, back-to-the-landers, 'deviants' of all kinds in all kinds of spaces. To the extent that these communities are the sources of energy ('difference') upon which postmodern states and corporations rely for their very existence, it could be said, as the autonomists say, that they have created the state and capital. But this process of co-optation, as they also note, is often contested, sometimes subverted, and never totally successful. This struggle defines the coming communities from another direction, as those identities that are not acceptable to, or at least not yet entirely normalized within, the global system. At their most radical limit, they present that which cannot be represented, that which *must not signify*—they are the disavowed, unconscious underside of globalizing capital, the Real that, just as it must be repressed, must just as surely return.

The disparities that allow the coming communities to act as crucibles for social change also mean that the simple dichotomy Agamben sets up between 'state' and 'humanity' is impossible to maintain. There need to be struggles not only between 'the state' (the bad?) and 'humanity' (the good?), but *within 'humanity' as well*. To postulate any identity category as unmarked and undifferentiated is, as I noted with respect to 'the multitude' and 'the proletariat', to assume a unity that not only must be striven for, but will never fully arrive. It appears possible only if one postulates a 'fundamental oppression', a substructure upon which all other oppressions are supposedly erected. As I pointed out in the previous chapter, however, this mode of analysis has become all but impossible to maintain. Pushing earlier critiques of the New Left to their logical and political limits, transnational feminists have convincingly argued that the great levelling and totalizing efforts of neoliberalism must be seen as just that—efforts, attempts, hegemonic constructs, fantasies imposed upon a field of endless and various struggle. As Chandra Mohanty points out:

The interwoven processes of sexism, racism, misogyny and heterosexism are an integral part of our social fabric, wherever in the world we happen to be. We need to be aware that these ideologies, in conjunction with the regressive politics of ethnic nationalism and capitalist consumerism, are differentially constitutive of all our lives in the early twenty-first century. (2003: 3)

While neoliberalism is globally present, and operates across all axes of domination and exploitation, we must keep in mind that it is manifested differently for different identities, at different times and places. A multidimensional analysis of oppression is therefore crucial to any effort to oppose, subvert or offer alternatives to the neoliberal world order.

A similar trajectory can be observed in debates within and around queer theory, which has been criticized for restricting its sphere to concerns that are 'merely cultural' (remember those new social movement theorists?). Judith Butler (1998) has responded by specifying the links between heterosexuality and capitalism, and Rosemary Hennessy (1996) has argued for a 'materialist' (that is, anti-capitalist) queer theory that gets past both modern essentialism (for example, in liberal multiculturalism) and postmodern anti-essentialism. A recently published collection entitled *Post-colonial, Queer: Theoretical Intersections* also contains articles that link a critique of heterosexism with a revolutionary materialist analysis of capitalist globalization (Morton 2001). This collection pushes the boundaries of the academic disciplines even further, by including writers who are critical of 'the heterosexist biases of postcolonial studies and the western biases of academic queer theory' (Spurlin 2000: 186). With these developments, queer theory is hardly restricting itself to cultural concerns, but is reaching towards a multidimensional analysis of global relations of power.

An equally important point, also raised by Mohanty, is the necessity of solidarity across these many dividing lines. Noting the many 'concrete effects', of global restructuring on women in various places and spaces, Mohanty argues for 'a more intimate, closer alliance between women's movements, feminist pedagogy, cross-cultural feminist theorizing, and these ongoing anti-capitalist movements' (2003: 245). On a similar note, but pushing the envelope beyond a specifically feminist solidarity, Beverly Baines, a Canadian anti-racist feminist activist, suggests that 'the frame we need to build ... is an integrative feminist anti-racist and *anti-oppression frame*' (cited in

Dua and Robertson 1999: 321; italics added). The key move here is contained in the reference to anti-oppression, which expands the field of anti-racist feminism to include a stand against hierarchical orderings as such. This could lead to what Sedef Arat-Koc has called a 'more engaged feminism', which would be interested in issues of equality and justice, both nationally and internationally, 'whether women may appear to be implicated in the issues or not' (2002: 63).

What, exactly, would this kind of feminism be like? One possibility is that it would be guided by what Gloria Anzaldúa calls a 'mestiza consciousness'. Living in between cultures and races, the result of a 'racial, ideological, cultural and biological cross-pollenization' (Anzaldúa 1987: 77), *la mestiza* has 'a plural personality' and is unable to 'hold concepts or ideas in rigid boundaries'. She is a creature for whom 'nothing is thrust out, the good the bad and the ugly, nothing rejected, nothing abandoned' (79). At first glance *mestiza* consciousness looks very much like the nomadic subject of postanarchism, which I have argued suffers from its claim to an impossible purity. But another aspect of *mestiza* consciousness is highlighted in Anzaldúa's contributions to *This Bridge Called My Back* (Moraga and Anzaldúa, 1981 2nd edn 1983) and *This Bridge Called Home* (2002). In these collections it becomes clear that Anzaldúa is more interested in *crossing* borders than she is in eliminating them, that she understands, because she lives, the dual dangers of integration and exclusion. In her preface to the second anthology, she notes that the first was limited to the voices of women of colour, many of whom viewed it as 'a safe space, as "home"'. But there are no safe spaces, Anzaldúa argues, adding that it is necessary to leave home and its illusion of safety if one hopes 'to bridge ... to attempt community' (2002a: 3).

Thus, in the follow-up collection there are articles from contributors who are not women of colour, who have been included in a self-conscious attempt to further the creation of what Anzaldúa calls 'El Mundo Zurdo', the left-handed world composed of those who 'do not fit':

Not all of us have the same oppressions ... we do not have the same ideology, nor do we derive similar solutions .... But these different affinities are not opposed to each other. In El Mundo Zurdo I with my own affinities and my people with theirs can live together and transform the planet. (2002b: 233)

The reference to 'transforming the planet' is an unfortunate Utopian echo that betrays the remnants of a hegemonic desire even in this explicitly affinity-based conception of social change. But Anzaldúa elsewhere shows that she is aware of the necessity of choosing when to cross borders, with whom and how to be open. 'Effective bridging', she warns, 'comes from knowing when to close ranks to those outside our home, group, community, nation—and when to keep the gates open' (2002a: 3). As an ethically committed subject, *la mestiza* necessarily abandons the position of pure nomadism—some things *are* thrust out, namely racism, sexism, homophobia ... perhaps capitalism and the state form as well. Thus, for example, despite its greater openness, there are no defenders of White male privilege in *This Bridge We Call Home*. Living affinity-based relationships means not only hooking up with those with whom we share values, but actively warding off and working against those whose practices perpetuate division, domination and exploitation.

Although feminists, postcolonial and queer theorists have all rejected totality in its various forms, there are currents within each of these traditions that help to guide us away from the trap of positing in its place a community that is entirely without presuppositions. Just as there can be no purely nomadic subject, there can be no purely nomadic community. There can, however, be communities that share presuppositions that are *different* from those of the global system of states and corporations, and that are at the same time changeable and open to *anything but* the emergence of apparatuses of division, capture, and exploitation. This is the crux of the task of building the coming communities: we must develop—and live according to—shared ethico-political commitments that allow us to achieve enough solidarity to effectively create sustainable alternatives to the neoliberal order. In the challenges to modernist liberation movements that have been advanced since the 1960s, it is possible to discern two intimately related themes that offer some guidance as to how these commitments might be understood.

THE COMING ETHICS:
GROUNDLESS SOLIDARITY AND INFINITE RESPONSIBILITY

Although it is clear that much work remains to be done, many first world feminists have responded favourably to the challenges posed by those who reject an identity-based politics of recognition, and their responses might be seen as showing the way forward for other

traditions. Donna Haraway's cyborg, for example, is a figure very
similar to *la mestiza*. S/he is a figure of 'liminal transformation' (1991:
177), part of a 'bastard race' that is 'stripped of identity' and 'teaches
about the power of the margins' (176). The cyborg is oriented to
crossing boundaries and overcoming dichotomies such as man/
woman, human/animal, organism/machine and material/virtual,
and participates in a non-hegemonic politics based on 'affinity, not
identity' (155). For many—myself included—it is hard to get into
the swing of Haraway's celebration of the radical possibilities of the
human–machine interface. As much as computerized systems are
dysfunctional and break down constantly, when they *do* work they
are perfect for implementing the kind of micromanagement desired
by the societies of control. Given the fact that very few human
subjects seem even momentarily to escape normalization, despite
being at least as unpredictable as computerized machines, it is hard to
imagine that cyborgs will fare any better. But then, as Haraway says,
the cyborg is a myth, and an ironic one at that. She is quite well aware
of the dangers of what she calls the informatics of domination, and
to the extent that she attempts to include the struggles of Southeast
Asian women working in electronic sweatshops in her vision, she at
least avoids the all too common trap of postulating a friction-free
virtual romp for 'everyone'—meaning everyone in the middle to
upper classes of the G8 countries, of course. Finally, although she is
out to revalue the highest values of western modernity, she is aware
of the danger of 'lapsing into boundless difference' (161), and thus
avoids the trap of pure nomadism as well.

Like Haraway, Braidotti also avoids postulating Woman as an
identity to be liberated and suggests that feminists must open
themselves to 'issues which at first sight seem to have nothing to do
specifically with women' at all. The qualifying phrase 'at first sight'
is crucial. What Braidotti is talking about is not an anything goes,
rent-a-radical form of activism; rather, she wants us to acknowledge
'the coexistence of feminine specificity with larger, less sex-specific
concerns'. What may not be identified as a 'women's issue' may in
fact be seen as having important repercussions for women, in which
case some feminists might find themselves interested in working
on it. Thus what she calls a 'nomadic feminism' is about 'tracing a
zigzagging path' (2002: 83) between the many issues faced by those
who are struggling to create alternatives to the neoliberal order.

While this formulation is clearly driven by an awareness of the
logic of affinity, the reference to 'larger' concerns sounds a little old-

leftish; and there are other signs that would indicate that Braidotti does not entirely reject the logic of hegemony. She suggests, for example, that 'real life minorities … women, Blacks, youth, post-colonial subjects, migrants, exiles and homeless may first need to go through a phase of "identity politics"—of claiming a fixed location' (84) before they begin to partake of 'newer' forms of political action. This formulation could be read as primitivizing and Eurocentric in its teleological connotations, and the idea that the thing for 'minorities' to do is to become 'major' enough to slough off their minority status is decidedly anti-Deleuzean in tenor. A politics of affinity, as Braidotti herself has argued, is not about abandoning identification as such; it is about abandoning the fantasy that *fixed, stable* identities are possible and desirable, that one identity is better than another, that superior identities deserve more of the good and less of the bad that a social order has to offer, and that the state form should act as the arbiter of who gets what. Although what constitutes a 'minority' at a particular place and time may change, and the composition of a given 'minority' identity will always shift, a poststructuralist analytics of power tells us that there will never be a time when there are no minorities. If we continue to put our energy into a politics of demand, we will be putting our energy into the perpetuation of the state form; we need to begin to understand better how, as persons of relative privilege, we can work to *demolish* our privilege without asking the state to do it for us.

A similarly ambivalent relation to the politics of demand can be found in the work of Diane Elam, who has argued for a feminism that 'would seek neither to liberate a female subject nor to secure certain fundamental rights for her' (1994: 77), but would, at the same time, continue to 'strategically appeal to rights politics' (80). Once again, I would suggest that state apparatuses are relatively uninterested in *why* we come to them to solve our problems for us. It is the fact that we come to them at all that perpetuates the state form as a state of relations in Landauer's sense. All the same, I find Elam's conception of a 'groundless solidarity' that is 'not based on identity, but on suspicion of identity' (69) to be very compelling. She arrives at this concept by asking herself a philosophical question about feminism and deconstruction, which she answers (partially) by suggesting that the relationship between them 'is not one of consensus (political common *ground*), but rather that of groundless solidarity' (25; italics in original). Elam is, of course, thinking of consensus in the Habermasian sense here, as something always achievable through the magic of language use, and therefore somehow always-already achieved. But at the same time she leads us towards a more anarchistic conception of consensus as a *process* rather than a state, something to be worked on via discussions that are tenuous and difficult. 'Groundless solidarity', she contends, 'is a stability but not an absolute one; it can be the object of conflict and need not mean consensus' (109). Thus we might finally have done with the idea that either conflict or consensus are 'at the heart' of human social relations; both are not only always possible, but always present, intermixing and at play.

Elam's intervention occurs in the context of the feminism/ postmodernism debates, but the concept of groundless solidarity has value for struggles on *all* axes of subordination, and especially for making links across these struggles. Trans theorist Leslie Feinberg has related how the early gay and lesbian movement suffered from the predominance of a 'similar middle-class White current' that had difficulty accepting the presence of drag queens at Pride events, and thereby 'weakened the movement they themselves depended upon for liberation' (1998: 58–9). Feinberg asks: 'What is the bedrock on which all of our diverse trans populations can build solidarity? The commitment to be the best fighters against each other's oppression' (60). The goal is not to 'strive to be one community' (*Sittlichkeit*), but to build many linked communities; not to 'find' leaders, but, as to recognize that everyone is a leader, that 'we are the ones we have been waiting for' (62).[1] That this potential is not merely theoretical is shown by the intense activity that is going on in activist circles around the world, to find ways to build concrete, practical links between disparate struggles, and to begin to engage in the extremely difficult task of dealing directly with the divisions that exist among us while resisting the temptation to pass this responsibility off to a state (or corporate) apparatus.

The No One Is Illegal campaign (NOII), operating out of Montreal, Quebec, is a loose coalition of activists who link neoliberal globalization to the displacement of people from the global South, who are compelled to leave their homes due to persecution, poverty or oppression. They often do so only to be categorized as 'illegal aliens' by the supposedly benevolent G8 countries where they seek refuge; they are denied the same rights as 'regular' citizens, and therefore face limited opportunities and further degradation. NOII also opposes the Canadian Aboriginal reserve system and state-based definitions of Aboriginal 'status', again linking the construction of these divisions to globalizing capital. NOII activists—who are affiliated with a broader campaign called Solidarity Across Borders—are involved in awareness-raising activities and direct action casework, and are committed to recognizing that 'struggles for self-determination and for the free movement of people against colonial exploitation are led by the communities who fight on the front lines' (NOII 2004). Although many of the people involved in NOII self-identify as anarchists, this identification is not central to their participation in Solidarity Without Borders, and usually goes unmarked. Their activities are carried out in a context of 'self-organized committees of persons directly affected by repressive anti-immigrant and anti-terrorist laws'—that is, NOII is directly involved in numerous activities of solidarity, without attempting to 'organize' anyone other than themselves. A similar outlook motivates the Indigenous Peoples Solidarity Movement and the Colombia Solidarity and Accompaniment Project, both based in Montreal, and Anti-Racist Action (ARA), which has chapters throughout Canada and the United States. These cells use direct action tactics to 'do the hard work necessary to decrease racism, sexism, anti-gay bigotry, oppression of religious freedom, and the unfairness which is often suffered by the disabled, the youngest, the oldest and the poorest of people' (ARA Montreal 2004). At the 2004 Montreal Anarchist Bookfair solidarity was a main theme, with a number of speakers from around the world urging attendees to join in struggles that may or may not be immediately identifiable as 'anarchist' in nature. This is, of course, not something particular to Montreal, or to Canada—these examples are a snapshot of a global tendency.

The model for this kind of solidarity work can be traced to the Zapatistas, who have been extremely effective in building world-wide support for their struggles against neoliberalism and the genocidal policies of the Mexican state. In an oft-cited interview in which his sexuality was brought into question, Subcomandante Insurgente Marcos, known to the popular imagination only as a vague, masked, figure, gave voice to the Zapatistas' vision of a global groundless solidarity:

> Yes, Marcos is gay. Marcos is gay in San Francisco, Black in South Africa, an Asian in Europe, a Chicano in San Isidro, an anarchist in Spain, a Palestinian in Israel, a Mayan Indian in the streets of San Cristobal, a Jew in Germany, a Gypsy in Poland, a Mohawk in Quebec, a pacifist in Bosnia, a single woman on the Metro at 10 pm, a peasant without land, a gang member in the slums, an unemployed worker, an unhappy student and, of course, a Zapatista in the

mountains. Marcos is all the exploited, marginalized, oppressed minorities resisting and saying 'Enough'. He is every minority who is now beginning to speak and every majority that must shut up and listen. He makes the good consciences of those in power uncomfortable—this is Marcos. (Marcos in People's Global Action 2002)

Many of the groups involved in mobilizing for the Battle of Seattle in 1999, such as Peoples' Global Action (PGA), Direct Action Network (DAN), and those who organized the first Independent Media Centre (IMC), had direct links with the Zapatistas (Callahan 2001: 38).

Among these, PGA is particularly interesting in terms of practices of solidarity, in that it was organized by ten activist groups from five different continents, including the Zapatistas, the Movimento dos Trabalhadores Rurais Sem Terra (MST, Brazilian Landless Peasants Movement), and Karnataka State Farmer's Union (KKRS), known for direct action against genetically engineered crops in southern India (PGA 2002). The group traces its roots to the *encuentros*, or meetings of solidarity, held in Chiapas in 1996 and Spain in 1997, and has since held global conferences in Geneva (1998), Bangalore (1999) and Cochabamba (2001). Its core principles, or hallmarks, include a 'very clear rejection of capitalism, imperialism and feudalism', as well as 'all trade agreements, institutions and governments that promote destructive globalization' (PGA 2004a, 1st hallmark). The rejection of feudalism may seem anachronistic to many readers; the group's website explains that it was added at the request of the Indian and Nepalese delegates, whose immediate struggles are oriented in this way. This is an excellent example of how Eurocentric assumptions about the nature of globalization can be challenged and the discourse of resistance enriched, given the right sort of organizational structure—one based on 'decentralization and autonomy' rather than hierarchical command (PGA 2004a, 5th hallmark). To maintain this structure, PGA has refused to take on a legal existence, and maintains that 'no organization or person represents the PGA, nor does the PGA represent any organization or person' (PGA 2004b). The group also maintains a multidimensional analysis of oppression within the neoliberal order, arguing that '[t]he denunciation of "free" trade without an analysis of patriarchy, racism, and processes of homogenization is a basic element of the discourse of the right' (PGA in Singh 2001: 49; cf. PGA 2004a, 2nd hallmark). Finally, it adopts what it calls a 'confrontational attitude', on the assumption that the politics of demand cannot have much effect on 'biased and undemocratic organizations in which transnational capital is the only real policy maker' (3rd hallmark). This attitude extends to support for, and participation in, direct action civil disobedience, and in the case of Chiapas, armed rebellion. There are many similar groups now in existence, such as Via Campesina, which is a network of peasant organizations, agricultural workers, rural women and indigenous communities from Asia, Africa, America and Europe. Like PGA, Via Campesina is organized along non-hierarchical lines and respects the local autonomy of its members.

While all of these organizations oppose the current organization of global capitalism and the statist institutions upon which it depends, they do not always challenge the neoliberal system of states as such. That is, while they are strong on building solidarity *against* the existing order, they are not as effective in building alternatives to it. Perhaps not surprisingly, given their hyper-exclusion from almost all of the so-called benefits of modernization and postmodernization, in combination with their precarious, yet vastly superior hold on traditional values and ways of life, it is indigenous peoples who are leading the way here.

Shunning the options of assimilation and integration within a White settler context, the indigenous peoples of mainland Australia have brought into being an Aboriginal Provisional Government (APG) that intends to achieve a total break with the Australian state. In choosing this path, the APG clearly abandons a politics of demand in favour of direct action:

> The APG anticipates small areas of land initially being given back to Aboriginal communities after specific campaigns over long periods of time. Political unification of those successful groups would form the developing Aboriginal nation territory. The strategy would be to rally all Aboriginal people around a particular community which is seeking to reclaim certain areas of land .... [C]ontrol would eventually be conceded by the white authorities as being revested in the Aboriginal communities. This of course would take great people resources, financial support, and grim determination. The latter is entirely in our hands. (APG 1992b)

Breaking with the Australian state, however, does not necessarily mean breaking with the state form as such. The APG uses a model in which it would become 'a nation exercising total jurisdiction over its communities to the exclusion of all others' (APG 1992a). The distinction here is literally Black and White: Aborigines inside their nation-state container are juxtaposed with all others, without any hint of linkages that might exist on either side of the Aboriginal/non-Aboriginal divide. Thus, although they reject the politics of demand and take up direct action, in seeking sovereignty on the European model this movement remains within the sphere of the hegemony of hegemony.

But a closer look at the nature of the internal relations advocated by the APG shows that the logic of affinity is also at work here. Well aware of the dangers of the liberal-capitalist societies of control, the APG papers acknowledge that 'there would be no point in transferring white power to an Aboriginal Provisional Government which simply imposed the same policies from above' (APG 1992c). Rather, they envisage a form of federation which gives maximum political control to local communities and minimizes the delegation of functionality and representation to the national level. Perhaps the most interesting aspect of this example is the assumption that there are not, nor should there be, enforced commonalities across Aboriginal communities at the level of social, political, and economic structures. The APG papers assume that some communities will tend towards relatively traditional systems, while others will opt for a mixture of Black and White law, and still others will choose a system which is 'simply appropriate to their life style in any given situation'. In vying for sovereignty on the European model as a means of protecting itself from the devastating ravages of the global system of states, the project of the APG displays a hybrid logic, combining elements of both hegemony and affinity.

A further example from the theory and practice of the indigenous peoples of North America will reveal more fully the avenues opened up by affinity-based practices. It is now well established that the model of the Haudenosaunee, or Iroquois Confederacy, was an inspiration for the federalists who founded the United States of America (Johansen 1999). But this model of inter-communal relations is not only of historical interest; it survives to this day and guides the political life of some of the most radical and self-reliant indigenous communities of North America. Its basic principle is given in what is known as the Two Row Wampum model, which is described in the following passage:

When the Haudenosaunee first came into contact with the European nations, treaties of peace and friendship were made. Each was symbolized by the Gus-Wen-Teh or Two Row Wampum. There is a bed of white wampum which symbolizes the purity of the agreement. There are two rows of purple, and those two rows have the spirits of your ancestors and mine. There are three beads of wampum separating the two rows and they symbolize peace, friendship, and respect.

These two rows will symbolize two paths or two vessels, traveling down the same rivers together. One, a birch bark canoe, will be for the Indian people, their laws, their customs and their ways. The other, a ship, will be for the white people and their laws, their customs and their ways. We shall each travel the river together, side by side, but in our own boat. *Neither of us will try to steer the other's vessel.* (Mitchell 1989: 109–10)

On this model, while there is a distinction between the 'Indian people' and the 'white people', this distinction is not cast in terms of an absolute dichotomy. In travelling the same rivers together, indigenous and non-indigenous peoples must be aware of their shared reliance upon the land and upon each other. But, in refraining from attempts to steer the other's vessel, each acknowledges the other's right to maintain its particularity and difference.

Recent texts in the field of Native American political theory have used this model as the basis for a critique of the integration of indigenous peoples within the neoliberal system of nation-states. As in western political theory, these critiques focus on issues of race, class, gender, and rational-bureaucratic domination of human beings and the land (Alfred 1999; Monture-Angus 1999; Maracle 1996; Marule 1984). Unlike most of their western counterparts, however, Native American political theorists also link these relations of subordination to the concept of sovereignty that serves as the horizon of the system of states itself. To place their critique in its proper context, it must be noted that the term sovereignty has many disparate meanings in discussions of indigenous self-determination. It can be invoked either as a justification for secession and the formation of a separate, independent state, as in the case of the APG, or it can serve as a basis for limited forms of self-government within an existing state, as in the official policies of the United States and Canada. Rejecting both of these models, the Mohawk nation of the Haudenosaunee have conceptualized a path of self-determination that involves neither a recovery of a partial remnant of a sovereignty lost in the past, nor a futural project of a totalizing nation-state. This approach is guided by the reflection that while redistribution of sovereignty may indeed challenge a particular colonial oppressor, it will not necessarily challenge the tools of his oppression. According to Taiaiake Alfred, sovereignty, as 'an exclusionary concept rooted in an adversarial and coercive western notion of power' is itself deeply problematic (Alfred 1999: 59). Taking up a position that is consonant with the Weberian critique of rationalization, Marie Smallface Marule and Lee Maracle argue that the structures and processes of bureaucracy that are necessary to postmodern sovereignty are oppressive as such, regardless of whether they are 'imposed' from 'outside', or 'chosen' from 'inside' a community (Marule 1984: 40; Maracle 1996: 52). Taken to its limit, this critique approaches that of the anarchist and anarchist-influenced groups described above, in positing modes of social organization in which there is 'no absolute authority, no coercive enforcement of decisions, no hierarchy, and no separate ruling entity' (Alfred 1999: 56).[2] The Haudenosaunee know what they want, because they've had it before; yet their struggle to walk their own path has been met with intense repression by all levels of the Canadian government, for which no tactic is too underhanded to ward off what it fears most—the successful construction of a form of indigenous governance that eludes its control, that does not rely in any way upon a devolution of 'authority' from the system of states. From the Oka Crisis of 1990 up to the present day, communities like Kahnesatake have been split internally and subject to constant external interference as they strive to find ways to maintain and reinvigorate their traditional forms of organization in the context of the postmodern condition.

Similar struggles have been faced by the Zapatistas, whose uprising against the Mexican state and the global neoliberal order has been accompanied by the emergence of an extremely interesting alternative system of social, political and economic organization. Although the army wing is organized hierarchically and is driven by authoritarian discipline, it is controlled by the General Council of the Clandestine Revolutionary Indigenous Committees, whose members are appointed by the local communities. These communities also may decide to become part of one of the 32 autonomous municipalities, which exist alongside and in struggle with the municipalities recognized by the Mexican state. This system, which encompasses a diversity of regions and ethnicities, arose as a direct action response to the perceived failures of the politics of demand:

> Before we would ask the government to give us everything, and they would give us handouts—some housing material, a little bit of money, a few sacks of corn. But now we realize that we can solve our necessities ourselves. That is

why we decided to resist, to give birth to our own ideas. The communities
created the autonomous municipalities so we could be free to create what
our thoughts tell us, to create what we want according to our needs and our
history. We are not asking the government to hand us clothes, but rather
the right to live with dignity. (Javier Ruiz in Chiapaslink 2000)

Within the communities decisions are made collectively on a
consensus model, in local assemblies where all men and women
over the age of twelve have a voice. Although they suffer from
constant persecution and have very little in the way of resources,
the communities and autonomous municipalities have managed to
make progress in areas ranging from health to education to economic
sustainability. Each community makes decisions for itself, pursues
the projects it wants to pursue, as does each municipality; as in
classical anarchist federalism, association does not mean giving
up local control. This is not to say, however, that the system of
autonomous municipalities is explicitly based on the ideas of Godwin
and Proudhon. Rather, like the nations of the Haudenosaunee, the
Zapatistas have carried out a creative application of traditional
indigenous structures in response to current exigencies. One
example is the way in which delegated authority is regulated so as
to ensure that those who are placed in positions of representation
serve their community rather than vice versa. 'When the authority
goes amiss, becomes corrupt or, to use a local term, "is a shirker",
he is removed from his position, and a new authority replaces him'
(Marcos 2003).

This system of recall is similar to that of the Mohawk nation of
the Iroquois Confederacy, except that in these communities it is
the women who select—and may depose—the men who will take
up positions of leadership (Arihwakehte 2004). In the indigenous
communities which gave birth to the Zapatista movement, it would
appear that women were not so empowered; patriarchal social
structures dictated that they were excluded from the decision-making
process. An important part of the Zapatista struggle has been to
address this disempowerment, through the inclusion of women
fighters in the EZLN and by challenging traditional structures
through the Revolutionary Law for Women. As a result of their own
autonomous struggles within the larger movement, women now
participate in the deliberations of the local communities and may be
sent as delegates to the CCRI and municipal councils (Millàn 1998:
67; Apreza 2003). Although some progress has been made, there is

also further work that needs to be done regarding the division of domestic/community labour, the valuation of women's work, and the subordination of women in domestic settings (Stephen 1995: 91; Millàn 1998: 74). Just as in the system of states, undoing patriarchy is a long and arduous task; what the Zapatista experiments show is that this kind of work can be done without a politics of recognition, that is, without an abstract political process through which women are 'given rights' by men as a favour or gift, under the coercive tutelage of a state apparatus.

THE PROBLEMS OF WHITE MIDDLE-CLASS MOVEMENTS

The struggles for autonomy being carried out by indigenous peoples around the world show that the coming communities are in fact beginning to arrive, that there is more to this notion than mere high theory. But, in observing that the logic of affinity guided by groundless solidarity does structure some actually existing movements and social experiments, I am not by any means trying to suggest that all of the work has been done. The problems encountered by women in the Zapatista autonomous zones are generalizable to other axes of oppression, other regions, movements, and traditions—every 'historical' society has been to some extent patriarchal, racist and homophobic. Judy Rebick, a Canadian feminist and anti-globalization activist, has recently pointed out that while the organizational methods of the anti-globalization movement have been profoundly influenced by feminist critiques of power, 'there is a very real risk that specific issues of concern to women such as male violence and reproductive rights are getting lost in the agenda' (Rebick 2002: 24). A similar point has recently been made by Chandra Mohanty (Dua and Trotz 2002: 67), and by many anarcha-feminist activists, who note that White male anarchists have a particularly strong tendency to consider themselves to have moved beyond the need to address how apparatuses of division affect their own practices:

> There seems to be an assumption at work that if we are fighting 'the system' that is oppressive, then we are somehow 'non-oppressive' by virtue of claiming to be 'outside' of the system. None of us are immune from the grasp of patriarchy, racism, and homophobia. The implications of thinking that we are immune can dangerously affect the participation of systematically oppressed peoples in the movement ... (Hewitt-White 2001: 20)

Numerous articles have been written on this point, the most widely circulated of them being Betita Martinez's 'Where was the Color in Seattle? Looking for reasons why the Great Battle was so White'. 'In the vast acreage of published analysis about the splendid victory over the World Trade Organization in 1999', Martinez wrote, 'it is almost impossible to find anyone wondering why the forty to fifty thousand demonstrators were overwhelmingly Anglo. How can that be, when the WTO's main victims around the world are people of color?' (Martinez 2002: 80).

In one of the interviews conducted for the Affinity Project, 'Africa'[3] a Black activist working in the Baltimore area, related his experience with people 'who still have a lot of conditioning they need to challenge ... and so maybe they make presumptions, maybe they have fears that are irrational' (Africa 2003). The examples he cites include a young woman at a Philadelphia squat who was surprised and confused to discover that he was an anarchist, 'just like her', and an IMC article that identified him as the lone African American rider at a Critical Mass demo. 'I didn't notice', Africa says, 'but it seemed important to them.' While acknowledging the suspicion that is encountered by White activists working in Black neighbourhoods, he is also frustrated by the assumption that he is the one to call upon if there is any work to be done in or with people of colour. 'The only thing that seems viable' for White activists, he suggests, is building trust by 'being a part of the community, making yourself available'. Clarita,[4] a Mexican organizer involved in helping to create collectives that give women the means for sustainable economic well-being, expressed similar frustrations. As a participant in an alternative education project set up by White US activists, she was horrified to find that what was supposed to be an anti-authoritarian project turned into what felt to her like another 'invasion': a small group of people 'basically controlled all the resources, everything that was going on in the camp ... it was paternalistic, hierarchical, authoritarian, racist, and sexist ... a total disaster' (Clarita 2003).

These activists echo voices that have been raised against racism and sexism in North American activist circles since at least the days of the US abolition debates (Farrow 2002: 15). Also, as the articles in the anthology *Gay Men and the History of the Political Left* (Hekma et al. 1995) attest, progressive social movements have also had trouble dealing with their gay, lesbian, bisexual and transgendered comrades. In an excellent paper on the work of French revolutionary activist and Daniel Guérin, who came out in 1965 after many years of suffering from his 'schizophrenic' existence as a gay anarchist-communist, David Berry comments on how things have changed, and how they remain the same:

> [Guérin's experience] doubtless reflects the endemic—if nowadays more carefully hidden—homophobia of the left and the labour movement; and also the persistent reluctance on the part of many historians of the left and of labour, even today and perhaps particularly in France (relative to, say, Britain and the US), to attach importance to forms of social inequality and oppression linked to gender and sexuality. (2003: 2)

Although it has not been the subject of as much discussion as racism and sexism—and precisely *because* of this lack—it is clear that homophobia is still a problem within contemporary radical social movements. Paul Matisz, a bisexual activist in Peterborough, Ontario, observes that there is a lot of anti-queer sentiment in the town at large, and confirms that he has encountered some 'passive homophobia' among activists as well. However, he has found that he is generally willing and able to help people 'unlearn the bad habits of straight society', and notes that the local activist community includes many openly GLBT individuals and is generally 'incredibly supportive' on queer issues. 'I don't think I would have been able to take such a strong stand on queer issues, or even come out myself', Matisz says, 'if I hadn't been surrounded by queer people, most of whom are not closeted' (Matisz 2003).

Finally, it should be pointed out that problems of division exist not only between those at the top of the hierarchies of privilege and 'everyone else'. They are also apparent between structurally disadvantaged communities themselves. A recent survey of 50 US activists on how the events of September 11, 2001 have affected their work shows some examples of the kinds of issues that can be encountered:

> September 11th has actually made it pretty clear that our organization [a multi-issue queer group] is not involved in the same communities as the white queer communities and made it clear who we want to be working with. It has strengthened [our] connections with low-income organizations and other social justice groups led by folks of color. (Luby 2001)

These subtle currents of affinity and disaffinity point to the need for an ethic of infinite responsibility that pushes the basis for groundless solidarity to ever-greater levels of complexity and commitment.

Like groundless solidarity, the concept of infinite responsibility comes from Derridean deconstruction and serves as its necessary complement. Simon Critchley, who has been very influential in working against the reading of Jacques Derrida as a postmodern relativist, argues that 'Derridean deconstruction has a horizon of responsibility or ethical significance, provided that ethics is understood in the Levinasian sense' (1999: 236). In this Levinasian sense, ethics takes on a meaning similar to that given in the ethics/morality distinction discussed in Chapter 5, in that it does not seek to impose any universal-normative procedures or codes (255). Rather, the 'face' of the other 'whom I cannot evade, comprehend, or kill' calls forth an infinite responsibility 'to justice, to justify myself' (5). Derridean/Levinasian ethics, then, relies upon the claim that the 'deep structure of subjective experience is always already engaged in a relation of responsibility or, better, responsivity to the other' (Critchley 1996: 33). Taking a similar line, but in an explicitly feminist register, Ewa Ziarek has recently argued that the politics of radical democracy 'cannot be based only on the hegemonic consolidation of dispersed struggles against manifold forms of oppression; rather, it has to be articulated in the gap between the ethos of becoming and the ethos of alterity, between the futural temporality of political praxis and the anarchic diachrony of obligation' (Ziarek 2001: 9). To put it in less jargon-laden terms, this means that as individuals, as groups, we can never allow ourselves to think that we are 'done', that we have identified all of the sites, structures and processes of oppression 'out there' and, most crucially, 'in here', inside our own individual and group identities. Infinite responsibility means always being ready to hear another other, a subject who by definition does not 'exist', indeed *must not* exist (be heard) if current relations of power are to be maintained. To respond means at least to have heard *something*—though one can never hear entirely 'correctly' or completely—and thus represents a crucial step on the way to avoiding the unconscious perpetuation of systems of division.

Once again, this is a question that is not merely 'academic'. As Lorenzo Komboa Ervin has argued, 'movements for social change in this 21st century will make a decisive mistake' if they ignore the specificity of struggles that are not directly oriented to state domination and capitalist exploitation. 'They will create a middle-class "white rights" movement which will not elevate the masses of the world's peoples' (Ervin 2001: 7). In the aftermath of the mass protest convergences of the late 1990s, and in response to day-to-day problems encountered in more locally-based work, activists of all persuasions have been ramping up their efforts to address the ways in which structured inequalities and exclusions divide anti-globalization and other struggles. In a proposal for a Zapatista-style *encuentro*, a coalition of IMC activists highlighted the issues that are at stake: 'how to build open, inclusive, decentralized structures of accountability, decision-making, and action locally, regionally, nationally, and globally?' How to 'bridge gaps in gender, colour, culture, age, access, language and "otherness" for capacity building and empowerment?' (IMC Encuentro Proposal Working Group 2000). Clearly, this is going to require much more than simply 'including' those who are 'excluded' by the invisible hierarchies inherent even in the most anti-authoritarian organizing styles. As Chris Crass has argued:

> The idea that we just need to get more people of colour to join our groups is an example of how white privilege operates. It carries the idea that we have the answers and how it just needs to be delivered to people of colour—as opposed to, people of colour have been organizing for a long time and we (white activists) have a lot to learn so maybe we could find a way to form alliances, relationships, and coalitions to work with folx of colour and be prepared to learn as well as share. (Crass n.d.)

Getting beyond a practice that looks all too much like state-based liberal multiculturalism will require White/male activists giving up control of movements, events and projects, listening rather than talking, linking up with existing organizations rather than duplicating, colonizing or depleting them because they do not seem to be guided by familiar models or led by familiar people. It will mean remembering that despite what may be a very real commitment to anti-oppression struggles, those of us who are privileged benefit from our positions in oppressive structures, primarily through not having to worry about the effects they have upon our own theory and practice. Infinite responsibility means being aware of this privilege and refusing/diffusing it to the greatest extent possible. More than anything, though, it means being willing to hear that you have not quite made it just yet, that you still have something more to learn.

While liberal theorists and politicians pretend that all is well—except for a few minor difficulties that can be overcome by making a new law, creating an NGO, or adjusting a bureaucratic function here and there—increasing numbers of people all over the world are converging on the notion that the new global order needs to be fought on all levels, in all localities, through multiple, disparate—yet interlocking—struggles. Although an immense amount of work remains to be done, it does seem that connections across the chasms created by apparatuses of division are being made, with increasing frequency and complexity. This gives us a reason to hope that groundless solidarity/infinite responsibility can offer an alternative to the politics of recognition and integration, that apparatuses of division can be significantly dismantled through direct, community-based action, rather than just being ameliorated—or even further entrenched—by state-based reforms. Groundless solidarity arises from a precarious 'unity in diversity' of its own, a complex set of (partially) shared experiences of what it means to live under neoliberal hegemony, what it means to fight it—and to create alternatives to it. It provides a basis for linking the coming communities, for creating relationships that do not divide us into disparate, defenceless subjects begging to be integrated by the dominant order.

7

Conclusion:
Utopian Socialism Again and Again

I am totally grateful to the sons of bitches who provoked
this whole thing. Because now the citizens will never be the
compliant lot that they were before. I tell you, I'm seventy
four years old, and I've seen a lot of wars, and from each
one I've come out better and stronger. Without crisis, you
just wallow along in the happiness of a fool. We need to
use crises to give a new birth to ourselves.

(Leonor, participant in Argentinian
neighbourhood assemblies)[1]

SIGNS OF FAILURE/SIGNS OF HOPE

The logic of hegemony has been exhausted. Marxist revolutions have
failed to achieve a transparent society and liberal reform has gone
neoliberal—that is, it has become reactionary rather than progressive
in tone. Even direct action to impede or ameliorate the advance of
the existing order is limited in its efficacy, as dead power eventually
flows around any obstacle put in its path. If it were impossible to
escape the logic of hegemony, this would certainly be a recipe for
despair. But as I have tried to show, there *are* other ways of achieving
social change. Poststructuralist theory teaches us that one of the
basic problems of contemporary politics is figuring out how to get
more people in more places to overcome not only their desire to
dominate others, but their own desire to be dominated as well. The
notion of singularity as an example suggests that the best way to do
this is to create sustainable alternatives to the existing order, to show
concretely that the way things are now is not the only way they can
be. As is so often the case, those on the margins are showing the way
that those at the centres must somehow learn to follow: *asambleistas*
in Argentina, LPM activists in South Africa, Zapatista villagers in
Chiapas, Mohawk warriors within/against North America, squatters
in London—all of these groups and movements are exploring the
possibilities of non-statist, non-capitalist, egalitarian modes of

social organization. They are working to reverse the colonization of everyday life by taking control over—and responsibility for—the conduct of their own affairs. Nothing is more important today than building, linking and defending autonomous communities of this sort. They offer the only hope of even partially escaping the disasters being brought upon all of us by the ongoing intensification of the worst effects of capitalism, the state form, racism, heterosexism and the domination of nature.

In this book I hope to have shown that much of this kind of activity is already going on. What signs are there, though, to indicate that it might accelerate? Will there continue to be gaps and margins in the neoliberal order, or will the societies of control be able to approximate more closely their goal of a perfect and total order? Without falling into the early twentieth-century marxist trap of relying upon historical 'necessity' rather than action, it is still possible to argue that the logic of capital itself can be an ally. For example, if we pull our heads out of the mass media for even a moment, it becomes obvious that the 'globalized' world of friction-free consumption covers only a very small portion of the planet, including a few dozen mega-cities and the transportation networks that connect them. Friction-free production, of course, exists nowhere, but the global division of labour and mass media self-censorship allow those living out the neoliberal fantasy to pretend that it does; *we* do not have to experience the assassination of labour leaders, the destruction of young women's bodies, the poisoning of entire cities and countries, that allow us to enjoy a throwaway life. But even this vision is too much under the sway of neoliberal ideology, for there is no large urban centre within any first-world enclave that does not contain areas of great deprivation. This is to say that the brave new world, just like the old world, is founded on exploitation, and therefore must create and amplify inequalities of all kinds. A capitalist system simply cannot be a totalized system, since there would be no one and nothing to exploit, no potential for *profit*. Globalization is in this sense a huge sham, akin to the signboards on the way to town and the pictures of gourmet meals in Terry Gilliam's film *Brazil*. For every island of apparent perfection that neoliberalism creates, there will arise more gaps, lacks and margins, more places that are not sufficiently profitable to locate production, too impoverished to provide markets for consumption, nor sufficiently 'beautiful' (i.e. untainted by capitalist exploitation!) to function as sites for the homes of media stars and the neoliberal elites.

Not only the logic of capital, but also the basic principles of engineering work against the fantasy of a totally integrated world. As any computer user knows, the more complex a system becomes the more difficult it is to maintain, the more often and catastrophically it fails. Stupidly, but fortunately for those of us who oppose the dominant order, the military/financial internet has been robbed of most of its distributed character, that is, of the diversity and decentralization that would have made it relatively stable and immune to attack despite its increasing complexity. Of course, what goes for the internet goes for all of the systems that make up the societies of control. So even as the neoliberal order advances, the weaker and more full of holes it becomes. There is no single global village—rather, countless borders within and between the territories claimed by the system of states emerge and subside in response to complex relations of power. It is in these gaps, on these margins, that spaces are available for experimentation: abandoned buildings, low-rent districts, rural areas with no obvious tourist potential, ghost towns and geographically inaccessible regions. These experiments may be carried out 'illegally', as in the Chiapas autonomous zones, squatted social centres and occupied factories, or they can clothe themselves in the attire of normality, as in co-ops and intentional communities. Each way of doing things has its own dangers: violent state repression, increasingly carried out with the help of corporate 'security firms', is commonly used to destroy experiments that can be seen as 'defying the rule of law', so that squatting, for example, paradoxically means being constantly on the run. On the other hand, those experiments that are less frightening to the existing order and have some legal basis for their existence must constantly battle with co-optation, which is generally the result of too much 'success'. (We can't work as a co-op anymore, we're just too big for all of that consensus-based, touchy-feely stuff!)

These are just two rather simple observations, but they point to a need for much more work to be done on theorizing the possibilities and pitfalls of constructing alternatives to the neoliberal order. This can only be a practical theory and a theoretical practice, one that avoids both the quiescence brought on by excessive abstraction *and* the frustrations inherent in setting out to 'do something' without paying adequate attention to what others are doing now and have done before. It requires the participation of activists who are aware of past and current debates within and across radical social movements, and of theorists who are willing to say more than 'the people will work out what is to be done'. This is perhaps not as tall an order as it might seem, since *activists are always already doing theory and theorists are always already political subjects*; the challenge lies in increasing our awareness and acceptance of this mutual implication, in finding more ways to explore more fully the tensions it creates and the possibilities it opens up. It is in this spirit that I offer, by way of a conclusion, a few observations on the challenges and opportunities faced by those who set out to achieve sustainable social change through affinity-based strategies and tactics. Along the way I will also address some common concerns about the viability of non-hegemonic social, political and economic forms.

The first point I want to make regards what it means to be a 'minority' in the Deleuzian sense. All too often it would seem that those of us who want to live differently feel a compelling need to register this difference within mainstream consciousness, so that even though we might achieve a fair degree of community and autonomy in a particular geographical space, such as a city neighbourhood or rural area, we find ourselves constantly orienting to the task of *converting* others. This is the hoped-for power of the example, which I would never deny, but it is also one of the traps of a hegemonic orientation that sees only two possibilities: being the ones 'on top', or one of many 'at the bottom'. Thinking outside of the logic of integration, being a minority does not appear as a reason to attempt a hegemonic reversal of one's relationship to the majority. Rather, it motivates the construction of spaces that are more fully minoritarian. Once again, the example of the Zapatista autonomous zones is interesting in that it represents a successful attempt to achieve a critical mass in a circumscribed geographical space, *in a way that is particular to that space*, to the subjects, structures, processes and histories that make it what it is. It is unlikely, for example, that sites for experimentation could be secured by armed rebellion in the global North, due to the much greater forces of violent repression under the command of these states and the long-standing docility of most of the population. At the same time, though, the Zapatista example is not *entirely* incommunicable—certain elements of this struggle can be creatively appropriated elsewhere.

The institutions of liberal-democratic nation-states, for example, can be used in ways that avoid the channelling of all political activity into the rights/recognition stream. As the participatory democracy experiments being conducted in Brazil show, it is possible to take advantage of systems that insist upon majority rule by tapping into, or creating, a critical mass in a limited space. Porto Alegre is one of five major cities that are governed by the socialist *Partido dos Trabalhadores* (Workers' Party), and has become famous for initiating a participatory budget programme based upon neighbourhood meetings that set priorities for capital expenditure on items such as sanitation, street paving and parks. Central to this approach is the commitment to giving priority to the *poorest* inhabitants rather than the richest. Since its inception in the 1990s the programme has been expanded to include the entire city budget, and councils have been created to discuss other issues, such as housing, health, culture and the environment. The idea has also spread, with varying degrees of success, to over 100 Brazilian cities (Abers 2002). Tarso Genro, the former mayor of Porto Alegre, sees participatory budgeting as part of the creation of a 'new non-state public space', which it undoubtedly is. However, this is a space based on what he calls a 'hegemonic consensus'; in typical Gramscian terms, it is in fact part of a 'new state with two spheres', one composed of 'managers' (the state) and the other of 'citizens' organizations' (civil society) (Genro 2003, Thesis 15).

A similar model is behind the experiment inaugurated in the state of Kerala, India in 1996. Here the opportunity for increased local control was created, paradoxically, by a nation-wide devolution of power under India's ninth Five Year Plan. The national government required all states to delegate certain of their functions to lower levels of government, but did not specify the way in which this had to be done. The socialist government in Kerala opted for a total inversion of the usual planning process and set out to create a bottom-up system based on democratic participation at the village and neighbourhood levels. This process proceeded over several years and involved meetings at various levels of delegation, facilitated by the training of thousands of activists. It culminated in the submission of over 150,000 project proposals, about half of which were funded (Franke and Chasin 1998). As in the Brazilian experiments, the participatory democratic process here existed only to produce suggestions for state action, rather than to find ways to proceed outside of the state's purview. Thus, in both of these cases, the degree of departure from hegemonic practices is small. But, since the work of defining priorities particular to various levels of interaction is what would have to be done in a truly non-statist federalism, these experiments do provide important guideposts as to what can be accomplished within and against liberal-democratic institutions.

The same kind of openings exist, and have been exploited, in some G8 countries. The Freetown of Christiana, in Copenhagen, is a long-running intentional community that has managed to survive since 1971 despite repeated police raids and hostility from various governments, developers and the more conservative of Danish citizens. The community began as an occupation of a disused military barracks, and grew into a base for Denmark's Free Hash movement. It is run with a minimal set of 'laws'—no hard drugs, no weapons, no violence and no trading within buildings or residential areas—and decisions are made on a consensus basis at open meetings. Unlike many much shorter-lived squats, Freetown has its own 'immigration policy' and has become a top attraction for tourists. In addition to the cafés and bars frequented by these visitors, the community has its own meeting hall, bath house, factories, smithies, postal service and radio station, and sponsors a free Christmas dinner as well as international conferences (Marshall 1999). Christiana is much more of a non-statist, non-corporate alternative than Porto Alegre or Kerala, while at the same time avoiding the greatest pitfall of intentional communities—isolation—through careful attention to its relation to 'the outside'. It is an excellent example of what can be achieved through the construction of sustainable minoritarian enclaves within and against the dominant order, even when one of the driving forces of the community (use of soft drugs) is subject to intense state repression.

Obviously, liberal institutions cannot be exploited where they do not exist in any meaningful way, as is the case in much of the global South and for many communities in countries of the global North. Private property, however, is alive and well all over the world, and is also vulnerable to careful subversion. Expropriation, as Kropotkin defined it, involves communities taking back resources that are held by individuals, states or corporations. This can be done by way of a violent revolution, of course, but the path of structural renewal suggests something different: pooling resources to build protective shells within which we might conduct our experiments with minimal interference and across which we might make links of solidarity. These spaces can be at least as autonomous as those inhabited by corporations, which have been managing to win rather a lot of 'freedom' for themselves lately. The logic is the same as that of using the liberal political system: majoritarian systems can be used to construct semi-permanent autonomous spaces where a critical mass of minoritarian subjects can congregate.

These are, of course, impure and risky strategies, but I would argue that *everything* is impure and risky, including doing nothing; so, the presence of impurity and risk are not in themselves reasons to dismiss any particular strategy. It is also hard to say exactly what the risks might be, given that so few experiments in non-hegemonic social change have ever been carried out. Intentional communities and back-to-the-land movements perhaps come closest to this model, but they have not always been as engaged as they could be with broader struggles.[2] From the point of view of the logic of affinity, communities that can delink without creating major effects of domination and exploitation should be welcome to do so. I raise this point because this is precisely the desire expressed by many of those who have suffered the most from Eurocolonial oppression, through currents such as US Black nationalism and traditional indigenous self-determination. Rejecting the logic of recognition and integration means accepting the need some communities have to heal themselves, rather than berating them for their 'illiberal' or 'statist' inclinations. As Lorenzo Komboa Ervin has argued, 'anarchists cannot take a rigid position against all forms of Black nationalism ... even if there are ideological differences about the way some of them are formed and operate' (1994: Part One). To take this line is to perpetuate White privilege rather than to challenge it. It also must be remembered, as Ashanti Alston points out, that 'nationalism and statism are different because nationalism can be anti-state' (1999). This distinction, of course, is also made by Native American political theory, and highlights the danger of critiquing Red and Black nationalisms without sufficient attention to the historical and theoretical specificity of their interventions.

At the same time, we cannot forget that global effects arise out of, and depend upon, micropractices of power. It would thus be analytically unsubtle—and politically unwise—simply to privilege the local over the global, as though *all* local practices will *necessarily* minimize domination. (Here it might be advisable to recall Laclau and Mouffe's insistence upon the contingent nature of the results of any movement, strategy or tactic.) Rather, it is important to bring into view practices of domination and resistance within the category of the local itself. It is also important to note that *purely* local strategies of resistance, that is, those that are unaware of, or insufficiently attentive to, their location within global strategies of domination, may be impotent. Resistance and the construction of alternatives must be networked, but then the question becomes: how are we to proceed so as to minimize the potential for linked local practices of resistance to become global practices of domination? Again, adherence to a purity principle could be fatal. Accepting the deep interconnectedness of the local and the global, the mainstream and the alternative, may allow us to create more spaces that are sustainable and at least *significantly different* from the great sameness that threatens to overwhelm us, rather than leaving us to wait for the perfect order that may arrive ... someday. The coming communities need not be—indeed *cannot* be—either entirely linked or delinked; rather, they will establish their own links in the process of building decentralized networks of alternatives.

The most pressing challenge faced by radical theoretical practice today, I would argue, is to understand better what the autonomists call exodus. There does need to be a leave-taking; there are many ways in which many of us are already taking lines of flight from the neoliberal order. But those who leave must have somewhere to go; as Deleuze and Guattari are careful to point out, no line of flight can be continued forever. The question, then, is: how it is possible re-territorialize sustainable alternatives rather than return to the status quo or slide out into individual psychosis/community self-destruction? Lines of flight passing through sustainable 'places' can be discerned in the dissemination of non-branded tactics such as RTS, FNB, IMC, social centre, and so on. But these non-branded tactics exist only parasitically, and therefore beg the question of how they will survive materially over the long term. That is, they beg the question of the coming economies. Clearly, new forms of community must involve the creation of alternatives not just in culture and polity, but in economic relationships as well.

WHAT OF THE COMING ECONOMIES?

Socialist and capitalist economies are hegemonic in that they seek to use a single mechanism to guide an entire system—state control in the former case, and a market free-for-all in the latter. All actually existing economic systems involve elements of both market freedom and state regulation, but what is crucial from the perspective of the logic of affinity is the *attempt* to create a monolithic field. Affinity-based economies do not display this kind of regularity; the only thing that can bind them is their shared commitment to minimizing domination and exploitation. But where, precisely, are the lines to be drawn between domination and non-domination, exploitation and non-exploitation? As I've mentioned, many classical anarchists came quite close to economic libertarianism by relying entirely upon contracts between individuals and small groups. But how much of what does an individual have to control before she becomes a capitalist? How big can a 'small business' get before it becomes a corporation? There is, of course, a vast amount of work that has been done on these questions, which for reasons of space I cannot address here. That will have to wait for another book, probably written by someone else, for just as I do not believe that the coming communities can be defined positively and totally, so I would suggest that the economies associated with them will also be diverse and multiple. There are, in fact, numerous experiments in non-capitalist/ non-state socialist economics being conducted, a few of which I'd like to mention, just to give a sense of the kind of possibilities that are being explored.

One of the best-known and more successful experiments in economic alternatives is the LETS system, which currently has 1,500 groups in 39 countries (Taris 2003). LETS is basically a community-based barter exchange, where goods and services are traded and debits/credits tracked by a local accounting system. It is a non-branded, decentralized network—anyone can start a LETS system and can link up with other systems. LETS users include individuals, groups and small businesses, for whom it functions as an alternative market. Clearly, LETS systems do not challenge the fundamental assumptions of capitalist market relations, nor do they develop a critique of the state form. They provide an alternative, certainly, but one that exists within the existing order, with no aspirations to drain or replace it.

Participatory Economics, or ParEcon, is explicitly linked to the marxist and anarchist traditions, and thus has a more radical approach. According to Tom Wetzel, ParEcon is 'an attempt to specify, in an economic program, what the necessary conditions are that would need to be achieved to have a sustainable economic system in which workers are no longer an exploited, subjugated class; that is, Participatory Economics is an attempt to specify the structure of a classless economic system, and thus an economic program for the 'self-emancipation of the working class' (Wetzel n.d.). Michael Albert, the leading proponent of ParEcon, says that it concerns itself with four key questions, to which it provides its own characteristic answers. To the question of how people should be paid for the work they do, ParEcon responds by arguing that remuneration should be based on 'effort and sacrifice, not profit, power or output' (Albert

2000: 3). Citing the anarcho-syndicalist Rudolf Rocker, Albert suggests that 'all people should have a say in decisions proportionate to the degree they are affected by them' (4). With regard to how workplaces should be organized, ParEcon advocates 'balanced job complexes' that would evenly distribute both enjoyable and arduous tasks. Finally, production and consumption decisions are to be made according to 'participatory allocation', that is, without either markets or central planning (6). In a recent talk at the New England Anarchist Bookfair (2001), Albert called ParEcon an 'anarchist economics', and he has suggested elsewhere that it is oriented to achieving what he calls 'non-reformist reforms' by 'building new institutions in the innards of the old' (2000: 159). In terms of both its goals and its means of achieving them, ParEcon would seem to be very much a project of structural renewal in the sense I have been using this term, right down to the experiments that have been conducted using this model. These include the radical publishing houses South End Press (Cambridge, USA), Z Magazine (Toronto) and Arbeiter Ring (Winnipeg, Canada), as well as the Mondragon Café, also in Winnipeg.

ParEcon, however, remains a Utopian project, in that like the alternatives proposed by Fourier and Owen, it begins with a body of theory that it seeks to implement in practice. Recent developments in Argentina, discussed in Chapter 1, offer a complementary model that begins with practice and is driven by the necessity of filling in the gaps left by the collapse of the national economy under the stresses of neoliberal restructuring. As Ezequiel Adamovsky recalls:

> When the first piquetero groups, barter markets, assemblies, and vision emerged, it was not the fruit of years of patient campaigning (there was almost no-one advocating these kinds of organizations before they were born), but a spontaneous, I would say intuitive creation. The whole economy and political system collapsed, the people did not trust any of the parties, leaders, or unions available, so they simply gathered with other people like themselves and asked each other 'Do you have any idea of what's going on here? What do we do to protect our lives?' (Adamovsky and Albert 2003)

However they may come to be, the power of alternative economies resides in the fact that every dollar we don't spend in the capitalist and state socialist economies weakens those economies. By making our own arrangements, we help to render hegemonic systems redundant and lessen our dependence upon them.

A UTOPIAN FUTURE: THE END OF 'CIVIL SOCIETY'

Taking a non-totalizing position with regard to economic activity also helps us to overcome the problems associated with theorizing 'the economy' as an autonomous 'sphere' of activity separated from other such 'spheres'. As I have shown in the chapters on the genealogy of hegemony, in liberal theory of both the capitalist and postmarxist varieties it is only when a civil society (which may or may not be identified with or be seen as including a capitalist market) is externally 'mediated' by a state form that the defining—and highly desirable— situation of 'pluralism' arises (Shalem and Bensusan 1994). From the liberal point of view, polities in which this distinction has been eliminated must become either 'totalitarian' (excessively ordered) or 'anarchic' (excessively disordered), depending upon whether it is the state or the civil society/economy that exceeds its proper boundaries. A similar perception exists in most brands of marxism, where state coercion is seen as an unfortunate, but necessary, evil on the way to a classless society. Within these paradigms, then, it is impossible to imagine that sufficient order can be achieved in (post)modern societies without recourse to the state form. Anarchists have advanced the heretical thesis that prior to the rise of the system of states, the so-called 'barbarians' did not live in a pre-social, Hobbesian state of perpetual war. Rather, as Kropotkin argued, village associations formed the basis of complex and mostly peaceful societies all over the world, and laid the foundations of what would come to be known as modern civilization. From this point of view, it is both possible *and* desirable for human beings to live without state intervention (political principle/hegemony), if sufficiently strong non-state and non-corporate modes of organization (social principle/affinity) exist to take on the tasks assigned to them in the other paradigms. This is to say that civil society, as we currently know it, is not only superfluous, but dangerous, as it presumes and reinforces both the state and corporate forms through the theory of autonomous spheres of activity.[3]

Giving up on 'civil society', of course, means giving up on 'the people', and this is a sore point for many whose conceptions of social change are guided by a hegemonic logic. This sentiment is revealed in the following passage from a recent book on radical democracy by Anna Marie Smith:

> The commitment to the promotion of democratic pluralism must entail the social obligation to construct the conditions in which self-determination for *everyone*—and especially for the traditionally disempowered—becomes possible. In the United States, for example, progress towards this goal could *only* be made after radical changes to the political system and massive redistributions in income, employment, access to education, and access to health care took place. (Smith 1998: 149)

Here we can see what should by now be familiar as a hybrid logic of reformist Revolution—incremental change to state-based institutions is expected to add up, via additive irradiation effects, to a on overall reconstitution of the target society. The insistence that everything must be changed at once for everyone arises from a powerful ideal of social justice: no one should be left behind as some of us march toward a better life. After a few hundred years of experience, however, it is starting to become obvious that what used to be called the masses have, in the G8 countries at least, imploded into what Jean Baudrillard calls the mass. 'Their strength is actual, in the present, and sufficient unto itself. It consists in their silence, in their capacity to absorb and neutralize, already superior to any power acting upon them' (Baudrillard 1983: 3). If Baudrillard is correct—and I think he is—no amount of irradiation can rid the social body of its cancerous tumours—the energy will simply be absorbed by the mass.

Baudrillard, of course, is widely criticized for the political nihilism to which his understanding of the silent majority leads him. But an acceptance of the limited prospects for revolutionizing the masses or reforming the mass in the G8 countries need not necessarily lead to this kind of conclusion. Baudrillard *had* to demand that we forget Foucault, because Foucault is among those who give us the resources to understood that Baudrillard's fatalism is the result of his inability to overcome the logic of hegemony. Revolution and reform have failed to produce the goods, it is true, and neither the masses nor the mass have any political potential. However, what it seems cannot *ever* be done *for anyone at all* using hegemonic methods can perhaps be done *by* some of us, *here and now*. This suggestion of course carries elitist overtones, which is why I have tried so hard to point out that it is not rich first-worlders who are leading the way, but those who are suffering the most from the depredations of globalizing capital. Following in their footsteps means that we of the global north must learn to meet our own needs locally, thereby limiting our participation in, and draining energy from, the neoliberal order. To the extent that we succeed in doing this, we undermine our privilege and stand in solidarity with those who do not share it. Also, by providing alternatives for those who can and will join the exodus from the neoliberal order, we open ourselves to sharing what we have built. In both the long and short terms, I would argue, more progressive social change can be achieved for more people by ridding ourselves of one of the final vestiges of the logic of hegemony, which is also traceable back to the Utopian socialists: the will to save everyone at once.

To avoid another sort of misreading, I want to make it clear that I am not advocating total rejection of reformist or revolutionary programs in all cases; to do so would be to attempt to hegemonize the field of social change. Rather, as will be obvious from a careful reading of the previous paragraph, I am citing what I see as the historically established *limited prospects* for these modes, and arguing that non-hegemonic strategies and tactics need to be explored *more fully* than has so far been the case. It's a matter of jettisoning some baggage that might allow us to range more widely, to search out the solution spaces more fully. In a time when we are surrounded by state and corporate forms, it is ridiculous to think that we can just ignore them; at the very least they need to be warded off, and one of the best ways of doing this is to put some energy into influencing how they evolve. At the same time, it should be obvious that I am privileging direct action for the construction of sustainable alternatives over revolution and reform. I would argue that *most* of our energy should go into the former, while we engage with the states and corporations only as necessary to further their slide into relative obscurity and to protect ourselves against their depredations. What is most compelling about structural renewal is its ability to achieve the goals of revolution and reform here and now, rather than putting them off to some distant place and time. And, in theory at least, if everyone joined the exodus at once, then the whole world *could* change in the way that those who believe in a simultaneous transformation desire. Indeed, this seems no less likely to occur via structural renewal than it does by way of the other methods.

One of the great advantages of the postmarxist theory of radical democracy was that it shifted theoretical attention from a millenarian proletarian revolution to the 'actually existing' new social movements and the citizens who compose them. It could be suggested that in advocating structural renewal by smiths operating outside of civil society, I am undoing this good work by regressing to a Utopian position. I hope that by providing a detailed genealogy of the theory and practice of structural renewal I have laid some of these concerns to rest. Like the statist new social movements before them, these non-statist experiments are actually existing, part of a historical trend, and need to be taken much more seriously in discussions of radical social change. One benefit of doing this is that we can begin to distinguish between what Deleuze and Guattari call radicle and radical forms of rhizomatic organization. Rhizomatic systems are 'acentred ... finite networks of automata in which communication runs from any neighbour to any other ... such that the local operations are coordinated and the final, global result synchronized without a central agency' (1987: 17). While the rhizome, as we have seen, is commonly associated with affinity-based, micropolitical modes of social change, is in no way an essentially 'good' alternative to a 'bad' tree. As is clear from the use of the internet both to develop a military-industrial complex (USA), and to organize against its excesses (Zapatistas), these forms are not mutually exclusive, but co-exist and mutate through ever-changing modes of employment and states of relative hegemony. Indeed, postmodern societies of control are becoming increasingly dependent upon decentred multiplicities that are, none the less, hierarchical and authoritarian in nature. It is crucial to mark the distinction Deleuze and Guattari make between these *radicle* rhizomatic forms that have significant arborescent effects, and those *radical* rhizomatic systems that are anti-hierarchical and preserve local autonomy to the greatest possible extent. Without keeping this distinction in mind, it would indeed be Utopian to believe, as many theorists and activists seem to, that *decentralization means autonomy*. It means no such thing, necessarily. Rather, decentralization just as easily, and much more likely under current conditions, means a shift from modern discipline to postmodern control. With the advent and rapid growth of radicle rhizomatic forms, maintaining this differentiation will become an ever-more pressing task, an analogue to the problem of positive feedback in Leftist strategies to use the state form as a disciplinary political tool.

I would also suggest that structural renewal based on the logic of affinity is less Utopian than either reform or revolution in its orientation to the realization of desired forms here and now. It is about building spaces, places or *topias* in the most literal sense of the term. It is also eminently practical in noting that these spaces can be found in the distopias and atopias that are being created alongside, and at a greater rate than, the neoliberal utopia of free-flowing capital. In its agonistic and co-operative engagements with other theoretical tendencies and social movements, anarchism has been shedding its modernist dependence upon a Utopian ideal and amplifying its commitment to making use of modes of organization that are already in existence. As Colin Ward notes, quoting Paul Goodman: 'A free society cannot be the substitution of a "new order" for the old order; it is the extension of spheres of free action until they make up most of social life' (1996: 11). With the corporations working to undermine the states, and both the states and corporations increasingly dependent upon a complex, rickety system of computerized control, there are now more gaps than ever to be exploited, more dis- and a-topias within which autonomous experiments might be conducted. Now, more than ever, it would seem that we are faced with a choice between anarchy and anarchism.

Notes

INTRODUCTION

1. Even though the APEC inquiry was, as many neoliberal commentators have pointed out, a waste of time and money, one cannot help but feel nostalgic about the days when police brutality and illegal arrests at protests were at least considered worth worrying about. Now, it seems, they are simply taken for granted.
2. Anzaldúa died in May 2004, just before she was to defend her dissertation at the University of California, Santa Cruz.

I. DOING IT YOURSELF:
DIRECT-ACTION CURRENTS IN CONTEMPORARY RADICAL ACTIVISM

1. I am aware that some will see me as imposing an identity where there may not be one, but I am also aware that it is impossible to do analysis without imposition, and I would plead that I am thinking in terms of partial identification rather than total identity. Also, while my own project is not hegemonic, it is certainly affinity-seeking; that is, I am interested in helping to make links across struggles that I hope will become less disparate. This is indeed a political act, it is about opinions and analysis, it is about relations of power, as both competition and co-operation.
2. The basic premises of the primitivist rejection of work are similar to those of autonomist marxism. See the discussion of the concept of *exodus* in Chapter 5.
3. The refusal of work has also been a theme of Italian autonomist marxism, particularly during the end of the 1970s, when the 'metropolitan Indians' were a major force. According to Franco Berardi (Bifo), they 'work only as much as is strictly necessary to buy the ticket for their next trip, live in collective houses, steal meat at supermarkets, and don't want anything to do with dedicating their lives to stressful, repetitive work which is, on top of it all, socially useless' (Berardi 1997: 137).
4. This is especially true with regard to the gratuitous use of images of partially clad young women. In one photograph, a reclining naked woman declares that 'The emancipation of the workers will be the work of the workers themselves' (*Internationale situationniste* no. 9, August 1964, p. 36). Similar examples can be found throughout the entire run of the journal.
5. See the CDC website at <http://www.geocities.com/billboardcorrections/index.htm>.
6. See the Surveilllance Camera Players performances listing at <http://www.notbored.org/scp-performances.html>.
7. See the Puppeteers' Cooperative Home Page <http://www.gis.net/Epuppetco/>.

8. There are many links between ELF, Earth First! and currents of anarchist activism oriented to ecological concerns. EF!'s website includes links to *Anarchy: A Journal of Desire Armed*, *Fifth Estate* (an anarchist journal with a strong primitivist bent) and *Green Anarchy*. Eco-action.org in the UK supports a website for the Anarchist Teapot, a collective that has run a number of squatted cafes in Brighton (<http://www.eco-action.org/teapot/index.html>).

9. The monopoly of the state is explicit, while the participation of the corporations in legitimate violence is best viewed as a kind of 'franchise agreement'.

10. For a practical account of the operation of affinity groups in the anti-globalization movement, see Unanimous Consensus (2001) or the ACT UP website at <http://www.actupny.org/documents/CDdocuments/Affinity.html>

11. See <www.indymedia.org> for a list of affiliated sites and for accounts of the genesis of some of the more well-known IMCs.

12. I would like to thank Enda Brophy for his work in locating and translating portions of the documents upon which this account is based. As yet, very little of the story of Italian social centres is available in English.

13. One could say that affinity-based politics do contain a universalizing moment that takes the form of a postmodernist performative contradiction. That is, they might be seen as motivated by a desire to *universalize an absence of universalizing moments*.

2. TRACKING THE HEGEMONY OF HEGEMONY:
CLASSICAL MARXISM AND LIBERALISM

1. While reference will be made to periods of time, I want to make it clear that the analysis I will present is not based upon mere novelty or simple succession. Rather, it relies upon the observation of shifting 'regularities in dispersion' (Foucault 1972: 38). Further, any shift in structures or relations that might be noted should not be read as implying that previous forms have been eradicated from the field. Rather, it is a matter of emergence and subsidence, of ever-changing manifestations of forces that can be seen as linked by lines of affiliation and as working in opposition to each other. Finally, it should be noted that in proceeding genealogically I make no claim to be producing *the* truth, or even *a* truth. Rather, I want to trace a line of descent that I find interesting and compelling due to my own ethico-political commitments and theoretical interests. Other genealogies are not only possible, they are necessary, and I welcome them.

2. It could be argued that the history of hegemony is bound up with a history of the state form that would take us back to the ancient empires that mark the beginning of what we know as 'civilization'. As important as it might be to trace the lineage back this far, limitations of space require that I begin the discussion when and where a cognate of the term hegemony first emerges.

3. See the *Compact Oxford English Dictionary* p. 753: from Greek *hegemonia*, from *hegemon* 'leader', from *hegeisthai* 'to lead'.

4. Examples here might include Ulrich Beck who, while he is more critical of neoliberal globalization than Held, also purports to reject the politics of Left and Right, and holds to a similar vision of 'cosmopolitan republicanism' (Beck 2000: 9). Perhaps the most notorious figure in this landscape is Anthony Giddens, author of *Beyond Left and Right* and architect of the 'third way'. Giddens' 'framework for a radical politics' is driven by a rights-based conception of 'universal human values' (1994: 20) that excludes a critique of the state form. He entirely ignores anarchist currents in political theory and practice by obstinately equating socialism with state-centric marxism (2), while presenting a vision of a 'generative politics' that draws heavily from anarchist traditions of mutual aid and decentralization (15). Giddens is perhaps *the* master theorist of domesticated, integrated radicalism.

5. Hegel's writings are notoriously slippery and difficult to characterize, and even more difficult to criticize, since he believes himself to be addressing and transcending all possible critiques as his argument proceeds. There is also a vast secondary literature on his work, which I will not be addressing in any systematic way. For these and other reasons, the treatment I am giving his work may appear superficial to academic specialists, but hopefully will remain accessible to non-specialists.

6. According to his doctrine, various other stages must be passed through, before self-consciousness arrives at the ultimate communion with Absolute Spirit.

7. This discussion touches only obliquely upon a set of debates that are crucial to western social and political philosophy, namely the interpretation of the 'master–slave dialectic' in the *Phenomenology of Spirit*. For further discussion, see Day (2000: 35–8).

8. Joseph Femia argues that, both pre- and post-prison, 'the essential structure of [Gramsci's] thought and the core of his political commitment was marxist and revolutionary—albeit innovative and flexible' (1981: 243). The only point of the war of position, Femia suggests, is to ensure the success of a full-frontal assault on state power. On this reading, which is supported by Massimo Salvadori, 'Gramsci's theory of hegemony is the highest and most complex expression of leninism' (1979: 252). Norberto Bobbio, on the other hand, argues that 'Gramsci's theory introduces a profound innovation with respect to the whole marxist tradition. *Civil society in Gramsci does not belong to the structural moment, but to the superstructural one*' (1979: 30; italics in original). Of course, Gramsci's texts are notoriously fragmentary, and the effects of prison censorship can be invoked to support or debunk almost any reading of the *Notebooks*.

9. I would go so far as to suggest that the concepts of 'false consciousness' and 'consciousness-raising' in classical marxism prefigure the consensual aspect of hegemony.

10. Lenin and Plekhanov, of course, went their separate ways at the time of the split between Bolsheviks and Mensheviks. However, both of these parties were committed to hegemonic forms of struggle.

11. For an in-depth discussion of the concept of civil society as it relates to Gramsci's theory of hegemony, see Bobbio (1979).

12. This is not to suggest, of course, that Gramsci accepts the teleological and hierarchical aspects of Hegel's account of the relations between these systems.

13. The friend/enemy distinction was developed in the context of postmarxist theories of hegemony by Laclau and Mouffe (1985).

3. TRACKING THE HEGEMONY OF HEGEMONY: POSTMARXISM AND THE NEW SOCIAL MOVEMENTS

1. Derrida develops this concept in *Spectres of Marx* (1994), where he is careful to distinguish it from Utopian and millennial orientations.

2. It should be noted that in choosing this particular text, I am operating on the assumption that while the theory of hegemony presented by this 'later' Laclau is more concise and subtly nuanced, it does not significantly diverge from that presented in his earlier collaborative work with Mouffe. Also, I see Laclau and Mouffe as having pursued different, but not incompatible trajectories since *Hegemony and Socialist Strategy*. Thus I would claim that what I am reading here is an up-to-date rendering of the current state of 'their' theory of hegemony, which avoids the problems associated with criticizing a text that is almost twenty years old as though it had been written today.

3. In *Hegemony and Socialist Strategy*, Laclau and Mouffe suggest that the 'new struggles ... should be understood from the double perspective of the transformation of social relations characteristic of the new hegemonic formation of the post-war period, and of the effects of the displacement into new areas of social life of the egalitarian imaginary constituted around liberal-democratic discourse' (1985: 165). My reading is that they believe that hegemony began to become possible with western modernity, and becomes in some sense mandatory with the advent of postmodernity.

4. One might say that modern nation-states have long 'known' this to be the case; but the logic of hegemony moves beyond this unconscious, fearful awareness by *acknowledging and celebrating*, rather than dissimulating, the impossibility of achieving a pure identity.

5. The impossibility of achieving a 'full' or 'complete' identity is crucial to Laclau and Mouffe's theory of articulation, and is based upon a Lacanian conception of the subject as lack. See Stavrakakis (1999) for further discussion of this point.

6. That is, the unfixity of the floating signifier arises from the contestation over meaning that occurs *between* competing discourses; that of the empty signifier is a result of its function as a general equivalent *within* a particular chain (see Laclau, in Butler, Laclau and Žižek 2000: 305).

7. In the United States, the debates centre on symbolic representations of internal Otherness, in educational curricula for example, and 'culture' is taken in a broad and ever-expanding sense, to include race, ethnicity, gender, sexuality, age, ability, and so on. In Canada, multiculturalism

is marketed as a salve to heal the wounds of a fragile federation formed through a history of violent conquest, and culture refers to an amalgam of racial-cultural-ethnic 'characteristics' based on the idea of the nation as a community of birth. In Europe, while internal diversity is also a concern of the established nation-states, there is the additional level of the EU to be considered. In this context, the culture in multiculturalism takes on much of the work of 'society', that is, it tends to describe an institutionally complete and relatively self-sufficient articulation of a population with a place and a bureaucratic apparatus.

8. For the European case, see Shore (1998). For Canada, see Day (2000).

9. A societal culture, according to Kymlicka, is based on a shared language, and 'provides access to meaningful ways of life across the full rage of human activities—social, educational, religious, recreational, economic—encompassing both public and private spheres'. See Kymlicka (1998: 27).

10. In dialogical conversations, 'utterly incompatible elements ... are distributed among several worlds and several full-fledged consciousnesses; they are presented not within one field of vision, but within several complete fields of vision of equal value'. See Bakhtin (1984: 12).

11. The most celebrated example of self-government achieved in recent Canadian history is the creation of the territory of Nunavut, in which the legislature enjoys a wide range of powers, including 'the administration of justice' (*Nunavut Act*, 40–41–42 Elizabeth II c. 23.1e), taxation (j), some aspects of land use and sale (i, r, s), education (m), and cultural and linguistic affairs (m, n). But the territory also has a 'Commissioner', appointed by the Governor in Council, who takes instructions from Ottawa (6.2) and sits in the legislative assembly, though s/he is not an elected representative. The Canadian state has also reserved the right to 'disallow any law made by the Legislature or any provision of any such law at any time within one year after its enactment' (28.2). Whatever powers the people and legislature of Nunavut have been granted, their use will be closely scrutinized, and if necessary overruled, by a colonial administration centred in Ottawa.

12. I have chosen this term because it is used by Taiaiake Alfred, one of the writers who includes himself within the tradition I am attempting to engage (Alfred 1999: xvi). As is the case with all discourses, the objects, boundaries, limits, and the name of 'Native American political theory' are ill-defined and highly contested.

13. This is not to suggest that radical feminism does not suffer from hegemonic overtones, as in Dworkin's desire for a women's revolution, or from acceptance of integration in certain aspects, as in Firestone's reproduction of the classical marxist faith in technology. Here I only want to note this discourse's contribution to a general anti-integrationist tendency.

14. In *The Metastases of Enjoyment*, Žižek situates the Kantian imperative as a command to 'renounce your desire, since it is not universalizable!' (1994: 69). In a note, he adds, 'I myself yielded to this temptation in the last chapter of *Looking Awry*, where I propose the maxim "do not violate the other's fantasy-space" as a complement to Lacan's ethics of persisting in

one's desire' (1994: 84, n. 18). The question of the status of these two moments in Žižek's work certainly deserves further discussion.

4. UTOPIAN SOCIALISM THEN ...

1. In this and all subsequent citations the pagination of the 1793 edition of the *Enquiry* will be used.
2. It is often argued that at the end of his life Saint-Simon became an advocate of the working class—a 'genuine' socialist (see Keith Taylor 1975: 48). But a careful reading of the *New Christianity* (Saint-Simon 1975/1825) shows that his references to improving the lot of the 'poorest and most numerous class' do not indicate a fundamental change in his politics. He still believed in the civilizing powers of Commerce (301) and held that 'great industrial establishments are doing more to improve the condition the poor class than any measure taken hitherto by the temporal or spiritual powers' (298–9). He still maintained that all should work towards 'the formation of that political system in which the general interests are managed by the most capable men [*sic*] in the sciences of observation, the fine arts, and industrial enterprises' (303). Thus, the rhetorical device of invoking the interests of the poor and most numerous classes seems to have been nothing more than an attempt to get the clergy on side with his existing programme, rather than a change in the nature of the programme itself.
3. Nicholas Rissanovsky echoes Buber's reading: 'Fourier paid constant, minute, even obsessive attention to arrangements in a phalanx, but he neglected the world outside it' (1969: 81). But it should be noted that some of Fourier's disciples discuss in detail the combination of phalanxes (unarchies) into duarchies and so on, eventually reaching a state where three duodecharchies (Europe/Africa, Asia/Oceana and North and South America) comprise a global system of Harmony (Godwin 1972/1844: 75).
4. Despite its importance to Bakunin's thought, the distinction between social and political revolution is not always respected. For example, Lewis Call says of Bakunin's *Statism and Anarchy*: 'This work encapsulates most of the major precepts and problems of what I call classical or orthodox anarchism. Like nineteenth century Marxism, Bakunin's anarchism attempts to bring about *a dramatic social and political revolution* in order to realize a utopian vision of total human emancipation' (2002: 67; italics added).
5. This makes it hard to accept the reading of those who, like Darrow Schechter, argue that Bakunin was 'aware of the need to develop alternative economies within capitalism until revolutionary consciousness spread to all oppressed social strata' (1994: 49). This turns Bakunin into a Saint-Simonian/Fourierian reformist, despite the former's many explicit denunciations of this aspect of the latter's programmes.
6. In his 1867 address to the League for Peace and Freedom, Bakunin notes that '[t]he defects of Saint-Simonianism are too obvious to need discussion. The twofold error of the Saint-Simonists consisted, first, in

their sincere belief that though their powers of persuasion and their pacific propaganda they would succeed in so touching the hearts of the rich that these would willingly give their surplus wealth to the phalansteries; and, secondly, in their belief that it was possible, theoretically, a priori, to construct a social paradise where all future humanity would come to rest' (Bakunin 1867: 116)

7. In a different version of this same piece, Bakunin is quoted as saying: 'In general, regulation was the common passion of all the socialists of the pre-1848 era, with one exception only. Cabet, Louis Blanc, the Fourierists, the Saint-Simonists, all were inspired by a passion for indoctrinating and organizing the future; they all were more or less authoritarians. The exception is Proudhon' (<http://www.blancmange.net/tmh/articles/reasprop.html> p. 116).

8. I am aware, of course, that Bakunin distances himself from Proudhon in other contexts, where he seems to accept Marx's critique in acknowledging that Proudhon 'remained an idealist and a metaphysician' throughout his career (1990/1873: 142).

9. It is interesting to note that Kropotkin also contributed to the (mis)reading of Proudhon as against centralization as such, rather than as advocating economic rather than political centralization. 'As to centralization and the cult of authority and discipline, which humanity owes to theocracy and to Imperial Roman law—all survivals of an obscure past—these survivals are still retained by many modern socialists, who consequently have not reached the level of their two predecessors, Godwin and Proudhon' (Kropotkin 1912: 14).

10. Landauer's forays into Jewish mysticism have had something to do with his marginal status, but it is perhaps more due to the lack of translations of his work—a lack that definitely needs to be addressed.

11. The best that relatively privileged subjects of the First World can do, as I will argue in Chapter 7, is to create alternatives that drain energy from the neoliberal project and thereby minimize the harm that it does, while at the same time working to build links of solidarity with those of the global North and South who are not part of the small, but highly privileged 'mass' that provides the ballast for the global ship of states.

5. ... AND NOW

1. This term came to me by way of an unpublished paper written by Enda Brophy, a PhD Candidate in Sociology at Queen's University, Kingston Ontario.

2. Before them, of course, Jürgen Habermas had described a different moment in the same relationship through his thesis of the colonization of the 'lifeworld' by the 'system' (1987: 301–73).

3. 'We are not anarchists', Hardt and Negri declare on p. 350 of *Empire*, 'but communists who have seen how much repression and destruction of humanity have been wrought by liberal and socialist big governments.' This is a curious disavowal, not only because it seems that they do protest

too much, but because of the odd exclusion of anarchism as a form of communism!

4. See Deleuze's *Foucault* (1988) for an extended discussion of Deleuze's relation to Foucault's work.

5. One must include oneself in critique where necessary!

6. See Harry Cleaver's genealogy in *Reading Capital Politically* for a full discussion of the roots of autonomist theory and practice (2000: 58–76).

7. In the autonomist lexicon, 'post-Fordist' seems to be in keeping with a postmodernist position that acknowledges changes in technologies and political spaces, but maintains a modernist commitment to the centrality of work and the working class.

8. The book from which this quote is taken is one of the first published by the social centre movement, and is a co-operative effort primarily of CSOA Leoncavallo (broadly autonomist) and Cox 18 (broadly anarchist) in which they survey the composition of the people frequenting their social centres and reflect at length on their histories, current conditions and future directions. I am indebted to Enda Brophy for carrying out the research on, and translation of, these sources.

9. That this understanding has failed to move sufficiently beyond the academy is the fault, I would say, of those academics of an activist orientation; hence this book, and particularly this chapter, which represents an attempt to work across the academic and activist worlds, to show how poststructuralist theory can and does inform struggles for radical social change.

10. May claims that this approach is 'not one ... that follows the quasi-transcendental path of Habermas and Apel' (May 1994: 141). This is because he believes that 'contingency and impurity do not form bars to ethical commitment', and that 'ethical commitments' can be found within, rather than 'beneath', discursive practice (146). However, inasmuch as May's position necessarily conflates the morality-ethics distinction, it fails to respect the crucial point of differentiation between discourse ethics and poststructuralism.

11. See my article 'Ethics, Affinity, and the Coming Communities' (Day 2001b) for a detailed discussion of the logic of affinity and poststructuralist ethics.

6. ETHICS, AFFINITY AND THE COMING COMMUNITIES

1. Feinberg attributes this excellent formulation to African-American poet June Jordan.

2. See Day (2001a; 2002) for a discussion of the possibilities of what some are calling 'anarcho-indigenism'.

3. This is a self-chosen pseudonym.

4. Clarita is a pseudonym chosen by the interviewer.

7. CONCLUSION: UTOPIAN SOCIALISM AGAIN AND AGAIN

1. In 'Argentines Speak Out: Voices from the Neighbourhood Assemblies' Available from: <http://argentinanow.tripod.com.ar/6.html> [accessed July 12, 2004].
2. For a taste of what intentional communities around the world are up to, see Bunker et al. (1997).
3. It should be noted that some proponents of ParEcon reproduce the liberal/ marxist paradigm of autonomous spheres. 'One mainstay of parecon is the relative separation of the political and economic spheres. It's assumed that certain affairs will be handled by political institutions, others by economic institutions' (Dominick n.d.).

References

Abers, Rebecca (2002) 'Daring Democracy: Porto Alegre, Brazil', in *New Internationalist* (December). Available at: http://www.thirdworldtraveler. com/South_America/Democracy_PortoAlegre.html [accessed, July 26, 2004]

Aboriginal Provisional Government (APG) (1992a) 'Intellectual Prisoners', *The APG Papers*. Available from: <www.faira.org.au/issues/apg01. html#prisoners> [accessed June 22, 2003]

—— (1992b) 'Law Reform and the Road to Independence', *The APG Papers*. Available from: <www.faira.org.au/issues/apg01.html#reform> [accessed June 22, 2003]

—— (1992c) 'Towards Aboriginal Sovereignty', *The APG Papers*. Available from: <www.faira.org.au/issues/apg01.html#sovereignty> [accessed June 22, 2003]

Ackelsberg, Martha (1991) *Free Women of Spain: Anarchism and the Struggle for the Emancipation of Women* (Bloomington: Indiana University Press)

Adamovsky, E. and Albert, M. (2003) 'Argentina and Parecon: Michael Albert Interviews Ezequiel Adamovsky', *Znet*. Available from: <http://www.zmag. org/content/print_article.cfm?itemID=3995&sectionID=41> [accessed April 18, 2003]

Adams, Jason (n.d.) 'Postanarchism in a Bombshell', *Aporia Journal*. Available from: <http://aporiajournal.tripod.com/postanarchism.htm> [accessed April 18, 2003]

Africa, Frank (2003) Interview with Sean Haberle, for the *Affinity Project*, Queen's University at Kingston, Ontario. Conducted July 16, 2003, in Baltimore USA

Agamben, G. (1993) *The Coming Community* (Minneapolis and London: University of Minnesota Press)

Ahooja, Sarita and Schmidt, Andrea (2003) 'Stop the Deportations: Solidarity With Montreal's Sans-Statuts', *New Socialist* (Jan./Feb.): 9–11

Ainger, Katherine (2002) 'Mujeres Creando: Bolivian Anarcha-Feminist Street Activists', in Dark Star (ed.) *Quiet Rumours: An Anarcha-Feminist Reader* (London: AK Press)

Albert, Michael (2000) *Moving Forward: Program for a Participatory Economy* (London: AK Press)

—— (2001) 'Anarchist Economics'. Talk given at the *New England Anarchist Bookfair*. Available from: <http://www.radio4all.net/proginfo.php?id=3952> [accessed January 28, 2002]

Alexander, M. Jacqui and Chandra T. Mohanty (eds.) (1997) *Feminist Genealogies, Colonial Legacies, Democratic Futures* (New York: Routledge)

Alfred, Taiaiake (1999) *Peace, Power, Righteousness: An Indigenous Manifesto* (Don Mills, Ontario: Oxford University Press)

Alston, Ashanti (1999) 'Beyond Nationalism But Not Without It'. Available from <http://www.illegalvoices.org/knowledge/writings_on_anarchism/ beyond_nationalism_but_not_without_it.html>. Accessed April 8, 2005

Anderson, Perry (1976) 'The Antinomies of Antonio Gramsci', *New Left Review* (Nov.–Dec.): 5–78

Anzaldúa, Gloria E. (1987) *Borderlands/La Frontera* (San Francisco: Spinsters/ Aunt Lute)

—— (2000/1982) 'Turning Points: An Interview with Linda Smuckler', in Ana Louise Keating (ed.) *Interviews/Entrevistas* (London: Routledge)

—— (2002a) Preface to *This Bridge Called Home* (London: Routledge)

—— (2002b) 'La Prieta', in *This Bridge Called Home* (London: Routledge)

Apena, Adeline (1995–6) 'Women's Cooperatives in Nigeria's Food Economy', *Africa Update Nigeria Revisited*, 3(1)

Apreza, Inés Castro (2003) 'Contemporary Women's Movements in Chiapas', in C. Eber and C. Kovic (eds.) *Women of Chiapas* (London: Routledge)

ARA (Anti-Racist Action) Montreal (2004) 'What is ARA?' (Montreal: ARA [pamphlet])

Arat-Koc, Sedef (2002) 'Imperial Wars or Benevolent Interventions? Reflections on "Global Feminism" Post September 11th', *Atlantis*, 26(2): 53–65

Arendt, Hannah (1977/1963) *On Revolution* (Harmondsworth: Penguin)

Ariwahkehte, Clifton (2004) 'Anarchism and Traditional Iroquois Culture'. Paper given at the Montreal Anarchist Bookfair, May 16, 2004

Aster, Consorzio, Centro sociale Cox 18, Centro Sociale, Leoncavallo, Primo Moroni (1996)' *Centri Sociali: Geographie Del Desiderio* (Milan: Shake)

Bagguley, Paul (1992) 'Social Change, the Middle Class and the Emergence of "New social movements": a critical analysis', *The Sociological Review*, 40(1): 26–48

Bahro, Rudolf (1986) 'Basic Positions of the Greens', in *Building the Green Movement* (Philadelphia: New Society Publishers)

Bakhtin, Mikhail (1984) *Problems of Dostoevsky's Poetics* (Minneapolis: University of Minnesota Press)

Bakunin, Mikhail (1990/1873) *Statism and Anarchy* (Cambridge: Cambridge University Press)

—— (1867) 'Federalism, Socialism, Anti-Theologism', Speech Delivered September 1867 to the League for Peace and Freedom. Available from: <http://www.blancmange.net/tmh/articles/reasprop.html> [Accessed July 28, 2004]

—— (1973/1866) 'Principles and Organization of the International Brotherhood', in A. Lehning (ed.) *Michael Bakunin: Selected Writings* (London: Jonathan Cape)

—— (1973/1870) 'Revolutionary Organization and the Secret Society', in A. Lehning (ed.) *Michael Bakunin: Selected Writings* (London: Jonathan Cape)

—— (1973/1895) 'On Federalism and Socialism', in A. Lehning (ed.) *Michael Bakunin: Selected Writings* (London: Jonathan Cape)

—— (1973/1871a) 'On Science and Authority', in A. Lehning (ed.) *Michael Bakunin: Selected Writings* (London: Jonathan Cape)

—— (1973/1871b) 'The Paris Commune and the Idea of the State', in A. Lehning (ed.) *Michael Bakunin: Selected Writings* (London: Jonathan Cape)

Barclay, Harold (1992) *People without Government: An Anthropology of Anarchy* (London: Kahn & Averill)

Battistuzzi, Dave (2003) Interviews with Sean Haberle, research assistant for the Affinity Project, Queen's University at Kingston, Ontario. Interviews conducted May 10 and July 19, 2003

Baudrillard, Jean (1983) *In the Shadow of the Silent Majorities: or, the End of the Social, and other Essays* (New York : Semiotext(e))

Beck, Ulrich (2000) *What is Globalization?* (Oxford: Polity Press)

Bell, John (1999) 'The End of Our Domestic Resurrection Circus: Bread and Puppet Theatre and Counterculture Performance in the 1990s', *TDR: The Drama Review* 43(3): 62–80

Benhabib, Seyla (1995) *Feminist Contentions: a Philosophical Exchange* (New York: Routledge)

Berardi, Franco (Bifo) (1998) *La Nefasta Utopia di Potere Operaio: Lavoro Tecnica Movimento nel Laboratorio Politico del Sessnatotto Italiano* (Roma: Cestelvecchi)

Berry, David (2003) 'Daniel Guerin's Engagement with "Sexology" from the 1950s and his Contribution to the Theorization of Sexuality and Gender from a Historical Materialist Perspective'. Presented at the conference *Socialism and Sexuality: Past and Present of Radical Sexual Politics*, Amsterdam, October 3–4, 2003

Bertram, Benjamin (1995) 'New Revolutions on the "Revolutionary" Politics of Ernesto Laclau and Chantal Mouffe', *Boundary* 2, 22(3): 81–110

Best, Steven and Douglas Kellner (1991) *Postmodern Theory: Critical Interrogations* (New York: The Guilford Press)

Bey, Hakim (1991a) 'Post-Anarchism Anarchy', in *TAZ. The Temporary Autonomous Zone, Ontological Anarchy, Poetic Terrorism* (New York: Autonomedia)

—— (1991b) 'The Temporary Autonomous Zone', in *TAZ. The Temporary Autonomous Zone, Ontological Anarchy, Poetic Terrorism* (New York: Autonomedia)

—— (1991c) 'The Lemonade Ocean & Modern Times: A Position Paper by Hakim Bey'. Available from: <http://www.spunk.org/texts/writers/bey/sp000917.html> [Accessed July 28, 2004]

—— (1991d) 'Black Crown and Black Rose: Anarcho-Monarchism and Anarcho-Mysticism', in *TAZ. The Temporary Autonomous Zone, Ontological Anarchy, Poetic Terrorism* (New York: Autonomedia)

—— (1993) *Permanent TAZs*. Available from: <http://www.left-bank.org/bey/permane2.htm> [accessed July 28, 2004]

—— (1996) *Millennium*. Available from: <http://www.left-bank.org/bey/millenni.htm> [accessed July 28, 2004]

Bhabha, H. (1994) *The Location of Culture* (London: Routledge)

Bhabha, H. and Comaroff, J. (2002) 'Speaking of Postcoloniality, in the Continuous Present: A Conversation', in David Theo Goldberg and Ato Quayson (eds.) *Relocating Postcolonialism* (Oxford: Blackwell)

Black, Bob (1985) *The Abolition of Work*. Available from: <http://wwww.t0.or.at/bobblack/abolishw.htm> [accessed May 18, 2002]

BLF (2004a) *The Art & Science of Billboard Improvement: a Comprehensive Guide to the Alteration of Outdoor Advertising*. Available from: <http://www.billboardliberation.com/guidebook.html> [accessed June 25, 2004]

—— (2004b) *The BLF Manifesto*. Available from: <http://www.billboardliberation. com/manifesto.html> [accessed June 25, 2004]

Bobbio, Norberto (1979) 'Gramsci and the Conception of Civil Society', in C. Mouffe (ed.) *Gramsci and Marxist Theory* (London: Routledge)

—— (1987) 'A Socialist Democracy?', in R. Griffin (trans.) *Which Socialism? Marxism, Socialism, and Democracy* (Minneapolis: University of Minnesota Press)

Boelscher-Ignace, M. and R. Ignace (1998) 'The Old Wolf in Sheep's Clothing? Canadian Aboriginal Peoples and Multiculturalism', in D. Haselbach (ed.) *Multiculturalism in a World of Leaking Boundaries* (Munster: LIT Verlag), pp. 133–54

Bookchin, Murray (1995) *Social Anarchism or Lifestyle Anarchism: An Unbridgeable Chasm* (London: AK Press)

—— (1998) *The Spanish Anarchists: The Heroic Years 1868–1936* (Edinburgh: AK Press)

Braidotti, Rosi (1997) 'Meta(l)morphoses', in *Theory, Culture & Society* 14(2): 67–80

—— (2002) *Metamorphoses* (Oxford: Polity Press)

Brophy, Enda and Coté, Mark (2003) 'Le Vite dei Manifestanti Infami', in *DeriveApprodi*, 25

Buber, Martin (1958/1949) *Paths in Utopia* (Boston: Beacon Press)

Buchanan, Ian (2002) 'What is "Anti-globalization?" I Prefer Not To Say', *Review of Education, Pedagogy, and Cultural Studies*, (24): 153–5

Bunker, Sarah et al. (eds.) (1997) *Diggers and Dreamers: The Guide to Communal Living 98/99* (London: Diggers and Dreamers Publications)

Butler, Judith (1990) *Gender Trouble: Feminism and the Subversion of Identity* (London: Routledge)

—— (1993a) 'Imitation and Gender Insubordination', in *Gay and Lesbian Studies Reader* (London: Routledge)

—— (1993b) *Bodies that Matter: On the Discursive Limits of 'Sex'* (London and New York: Routledge)

—— (1998) 'Merely Cultural', *New Left Review*, 227 (Jan./Feb.): pp. 33–44

Butler, J., E. Laclau and S. Žižek (2000) *Contingency, Hegemony, Universality: Contemporary Dialogues on the Left* (London and New York: Verso)

Call, Lewis (2002) *Postmodern Anarchism* (Lanham: Lexington Books)

Callahan, Manuel (2001) 'Zapatismo and the Politics of Solidarity', in Eddie Yuen et al. (eds.) *The Battle of Seattle: The New Challenge to Capitalist Globalization* (New York: Soft Skull Press)

Carr, Marilyn, Martha Chen and Renana Jhabvala (eds.) (1997) *Speaking Out: Women's Economic Empowerment in South Asia* (Dhaka: University Press Limited)

Chiapaslink (2000) *The Zapatistas: A Rough Guide* (Bristol: Chiapaslink)

Cimino, Roberto (1989) *Testimonianza Di Roberto Cimino ('75–'78)*. Available from: <http://www.ecn.org/leoncavallo/storic/cimino.htm> [accessed May 16, 2002]

Clarita (2003) Interview with Sean Haberle, for the *Affinity Project*, Queen's University at Kingston, Ontario. Conducted July 20, 2003

Clarke, John (2001) *Short History of OCAP*. Available from: <http://www.ocap. ca/archive/short_history_of_ocap.html> [accessed May 16, 2002]

Clastres, P. (1989) *Society against the State* (New York: Zone Books)

Cleaver, Harry (1994) 'Kropotkin, Self-valorization, and the Crisis of Marxism', *Anarchist Studies*, 2: 119–36

—— (1998) 'The Zapatistas and the Electronic Fabric of Struggle', in J. Holloway and E. Peláez (eds.) *Zapatista: Reinventing Revolution in Mexico* (London: Pluto Press)

—— (2000) *Reading Capital Politically* (New York: Semiotext(e))

Collins, Patricia Hill (1991) *Black Feminist Thought: Knowledge, Consciousness, and the Politics of Empowerment* (New York: Routledge)

Committee for the Real Estate Show (1980) *The Real Estate Show Manifesto or Statement of Intent*. Available from: <http://www.abcnorio.org/about/history/res_manifesto.html> [accessed May 23, 2004]

Coté, Mark (2003) 'The Italian Foucault: Subjectivity, Valorization, Autonomia', *Politics and Culture*, 3

Crass, Chris (n.d.) 'Confronting Global Capitalism and Challenging White Supremacy: Thoughts on Movement Building and Anti-Racist Organizing'. Available from: <http://www.staralliance.org/starcives/cws.html> [accessed May 23, 2004]

Critchley, S. (1996) 'Deconstruction and Pragmatism—Is Derrida a Private Ironist or a Public Liberal?', in Chantal Mouffe (ed.) *Deconstruction and Pragmatism* (London: Routledge)

—— (1999) *The Ethics of Deconstruction*, 2nd edn (Edinburgh: Edinburgh University Press)

Critical Arts Ensemble (CAE) (1994) *The Electronic Disturbance* (New York: Semiotext(e))

—— (1996) *Electronic Civil Disobedience* (New York: Semiotext(e))

Crossley, Nick (2003) *Making Sense of Social Movements*. Buckingham: Open University Press

CSA (2003) Anonymous interview with former Italian social centre activist. Conducted July 2003 by Enda Brophy

Dalla Costa, Maria Rosa and Selma James (1973) *The Power of Women and Subversion of the Community* (Bristol: Falling Wall Press)

Day, Richard J.F. with T. Sadik, (2000) *Multiculturalism and the History of Canadian Diversity* (Toronto: University of Toronto Press)

—— (2001a) 'Who is This We that Gives the Gift? Native American Political Theory and the Western Tradition', *Critical Horizons*, 2(2): 173–201

—— (2001b) 'Ethics, Affinity, and the Coming Communities', *Philosophy and Social Criticism*, 27(1): 21–38

—— (2002) 'BC Land Claims, Liberal Multiculturalism, and the Specter of Aboriginal Nationhood', *BC Studies* (Summer): 5–34

Deleuze, Gilles (1992) 'Postscript on the Societies of Control', *October* # 59 (Winter): 3–7

—— (1995) *Negotiations (*New York: Columbia University Press)

—— (1998) *Foucault* (Minneapolis: University of Minnesota Press)

Deleuze, G. and F. Guattari (1987) *A Thousand Plateaus: Capitalism and Schizophrenia* (Minneapolis: University of Minnesota Press)

Deleuze, G. and Parnet, C. (1983) 'Politics', in John Johnston (trans.) *On the Line* (New York: Semiotext(e))

Deloria Jr., Vine and R.M. Lytle (1984) *The Nations Within: The Past and Future of American Indian Sovereignty* (Austin: University of Texas Press)

Del Re, Alisa (1996) 'Women and Welfare: Where is Jocasta?', in M. Hardt and P. Virno (eds.) *Radical Thought in Italy* (Minneapolis: University of Minnesota Press), pp. 98–113

Derrida, J. (1978) 'Structure, Sign and Play in the Discourse of the Human Sciences', in *Writing and Difference* (Chicago: University of Chicago Press), pp. 278–94

—— (1994) *Spectres of Marx* (London: Routledge)

Derrida, J. with Kristine McKenna (2002) 'The Three Ages of Jacques Derrida: An Interview with the Father of Deconstructionism', *LA Weekly* (November): 8–14

DIAND (Department of Indian Affairs and Northern Development, Canada) (1997) *Gathering Strength: Canada's Aboriginal Action Plan* (Ottawa: Minister of Public Works and Government Services)

Dogra, Bharat (2002) *Whither the Chipko Years? The Fading Gains of Himalayan Conservation.* Available from: <http://www.indiatogether.org/environment/articles/postchipko.htm> [accessed August 13, 2002]

Doherty, H. (1968/1851) Introduction to *The Passions of the Human Soul and Their Influence on Society and Civilization*, vol. 1 (New York: Augustus M. Kelley)

Dominick, Brian (n.d.) *ParEcon, Anarchy and Politics.* Participatory Economics Project. Available from: <http://www.parecon.org/writings/brian_state.htm> [accessed August 13, 2002]

Dua, Ena and Angela Robertson (1999) *Scratching the Surface: Canadian Anti-Racist Feminist Thought* (Toronto: Women's Press)

Dua, Ena and Alissa Trotz (2002) 'Transnational Pedagogy: Doing Political Work in Women's Studies. An Interview With Chandra Mohanty', *Atlantis*, 26(2): 66–77

Dworkin, Andrea (1974) *Woman Hating* (New York: Dutton)

—— (1989) *Letters from a War Zone: Writings, 1976–1989* (New York: E. P. Dutton)

Dyer-Witheford, Nick (1999) *Cyber-Marx: Cycles and Circuits of Struggle in High-Technology Capitalism* (Urbana and Chicago: University of Illinois Press)

—— (2002) 'Global Body, Global Brain/Global Factory, Global War: Revolt of the Value Subjects', *The Commoner*, 3 (January): 1–30

Dykstra, Mathieu (2003) Interview with Sean Haberle, for the *Affinity Project*, Queen's University at Kingston, Ontario. Conducted May 10, 2003 in Montreal, Quebec

Earth First! (2004) *Earth First in Britain.* Available from: <http://www.earthfirst.org.uk> [accessed September 11, 2004]

Eisenstein, Z. (1979) 'Developing a Theory of Capitalist Patriarchy and Socialist Feminism' in *Capitalist Patriarchy and the Case for Socialist Feminism* (New York and London: Monthly Review Press), pp. 5–40

Elam, Diane (1994) *Feminism and Deconstruction: Ms. en abyme* (London: Routledge)

Eldridge, Dan (2003) 'Anarchy in the Steel City', *Pulp* (May 29, 2003–June 5, 2003)

Eleanor (2003) Interview with Sean Haberle, research assistant for the *Affinity Project*, Queen's University at Kingston, Ontario. Interview conducted May 16, 2003, at Concordia University, Montreal, Quebec

ELF (2004) *Meet the E.L.F.* Available from: <http://www.earthliberationfront. com/about> [Accessed September 11, 2004]

Eliot, T.S. (1944) 'East Coker', in *Four Quartets* (London: Faber and Faber)

Engels, Friedrich (1978/1872) 'Versus the Anarchists', in Robert Tucker (ed.) *The Marx–Engels Reader* (New York: Norton)

—— (1978/1880) 'Socialism: Utopian and Scientific', in Robert Tucker (ed.) *The Marx–Engels Reader* (New York: Norton)

—— (1982/1845) 'Engels to Marx in Brussels', in *Collected Works of Marx and Engels, Vol. 38: Correspondence 1844–1951* (New York: International Publishers)

Engler, Yves and Bianca Mugyenyi (2004) 'The Take: Factory Occupations', in *Znet Daily Commentary*. Available from: <http://www.zmag.org/sustainers/ content/2004–06/11engler-mugyenyi.cfm> [accessed May 23, 2004]

Ervin, Lorenzo Komboa (1994) *Anarchism and the Black Revolution*. Available from: <http://www.illegalvoices.org/apoc/books/abr/> [accessed August 18, 2004]

—— (2001) *Colours of Resistance* #1 (Spring)

Fanon, Frantz (1963) *The Wretched of the Earth* (New York: Grove Press)

Farrow, Lynne (2002) 'Feminism as Anarchism', in Dark Star (ed.) *Quiet Rumours: An Anarcha-Feminist Reader* (London: AK Press)

Feinberg, Leslie (1998) *Trans Liberation: Beyond Pink or Blue* (Boston: Beacon Press)

Femia, Joseph V. (1981) *Gramsci's Political Thought: Hegemony, Consciousness, and the Revolutionary Process* (Oxford: Clarendon)

Firestone, S. (1970) *The Dialectic of Sex: The Case for Feminist Revolution* (New York: William Morrow)

Flood, Andrew (2002) 'A Review of Negri and Hardt's *Empire* from an Anarchist Perspective: Is the Emperor Wearing Clothes?', *DIY Publishing*. Available from: <http://struggle.ws/pdf/empire.html> [accessed May 23, 2004]

Food Not Bombs (2004) *The Food Not Bombs Story*. Available from <http://www. foodnotbombs.net/> [accessed August 5, 2004]

Foucault, Michel (1972) *The Archaeology of Knowledge* (London: Routledge)

—— (1979) *Discipline and Punish* (New York: Vintage Books)

—— (1980) 'Gouvernement des Vivants'. Lecture delivered January 30 at Collège de France

—— (1985) 'Nietzsche, Genealogy, History', in P. Rabinow (ed.) *The Foucault Reader* (New York: Pantheon)

—— (1987) 'The Ethic of Care for the Self as a Practice of Freedom', *Philosophy and Social Criticism*, 12 (2–3) (Summer)

—— (1990) *The History of Sexuality*, vol. 1 (New York: Vintage Books)

—— (1991) 'Governmentality', in Graham Burchell, Colin Gordon and Pete Miller (eds.) *The Foucault Effect: Studies in Governmentality, with Two Lectures by and an Interview with Michel Foucault* (London: Harvester Wheatsheaf), pp. 87–104

—— (1997a) 'What is Enlightenment?', in P. Rabinow (ed.) *Michel Foucault: Ethics, Subjectivity and Truth* (New York: New Press), pp. 303–20

—— (1997b) 'Polemics, Politics, and Problematizations', in P. Rabinow (ed.) *Michel Foucault: Ethics: Subjectivity and Truth* (New York: New Press), pp. 303–20

—— (2003/1976) '17 March 1976', in *Society must be Defended: Lectures at the Collège de France 1975–1976* (New York: Picador), pp. 239–64

Foucault, M. and Deleuze, G. (1976) 'Intellectuals and Power', in *Foucault Live*, (New York: Semiotext(e))

Fourier, Charles (1968/1951) *The Passions of the Human Soul* (New York: Augustus M. Kelley)

—— (1971a) *Design for Utopia: Selected Writings of Charles Fourier* (New York: Schocken Books)

—— (1971b) 'Studies in Psychopathology', in J. Beecher and R. Parvenu (eds.) *The Utopian Vision of Charles Fourier* (Boston: Beacon Press)

—— (1996/1808) *The Theory of the Four Movements* (Cambridge: Cambridge University Press)

Franke, Richard and Chasin, Barbara (1998) 'Power to the (Malayalee) People: Democracy in Kerala State, India', *Z Magazine* (February). Available from: <http://www.thirdworldtraveler.com/Democracy/Malayalee_KeralaZ.html> [accessed August 4, 2004]

Fraser, N. (1981) 'Foucault on Modern Power: Empirical Insights and Normative Confusions', *Praxis International* 1(3): 272–87

—— (1997) 'Heterosexism, Misrecognition, and Capitalism: A Response to Judith Butler', *Social Text* 52/53, 15 (3 & 4) (Fall/Winter): 279–89

Fukuyama, Francis (1992) *The End of History and the Last Man* (New York: Free Press)

Gambone, L. (1996) *Proudhon and Anarchism: Proudhon's Libertarian Thought and the Anarchist Movement* (Montreal: Red Lion Press)

Garrigues, Lisa (2002) 'Politics without Politicians: An Update on the Argentine Assemblies'. Available from: <http://www.argentina.indymedia.org/news/2002/06/29714.php> [accessed July 15, 2004]

Genro, T. (2003) 'From Brazil to the World, or Twenty Theses for a Democratic Theory of the State', 11 July. Available from <http://www.opendemocracy.net/articles/newPopUpArticle.jsp?id=3&articleId=1355>. Accessed April 8, 2005

Geras, Norman (1987) 'Post-Marxism?', *New Left Review* (163): 40–83

Giddens, Anthony (1994) *Beyond Left and Right: The Future of Radical Politics* (Cambridge: Polity Press)

Godwin, Parke (1972/1844) *A Popular View of the Doctrines of Charles Fourier* (Philadelphia: Porcupine Press)

Godwin, William (1993/1793) *An Enquiry Concerning Political Justice: Political and Philosophical Writings of William Godwin*, vol. 3 (London: William Pickering)

Graeber, David (2002) 'The New Anarchists', *New Left Review* (13): 61–73

Gramsci, Antonio (1971) *The Prison Notebooks*, ed. and trans. Q. Hoare and G. Smith (New York: International Publishers)

Greenberg, Stephen (2004) 'Post-apartheid Development, Landlessness and the Reproduction of Exclusion in South Africa' (Durban: Centre for Civil Society Research). Available from: <http://www.landaction.org/display.php?article=232> [accessed August 15, 2004]

Grossberg, Lawrence (1992) *We Gotta Get Out of This Place* (New York and London: Routledge)

Guattari, F. (1995) *Chaosophy* (New York: Semiotext(e))

Guattari, F. and Negri, A. (1990) *Communists Like Us* (New York: Semiotext(e))

Guilloud, Stephanie (2001) 'Spark, Fire, and Burning Coals: An Organizer's History of Seattle', in E. Yuen et al. (eds.) *The Battle of Seattle: The New Challenge to Capitalist Globalization* (New York: Soft Skull Press)

Habermas, J. (1984) *The Theory of Communicative Action* (Boston: Beacon Press)

—— (1987) *The Philosophical Discourse of Modernity* (Cambridge, Mass: MIT Press)

Hall, Stuart (1983) 'The Great Moving Right Show', in S. Hall and M. Jacques (eds.) *The Politics of Thatcherism* (London: Lawrence & Wishart), pp. 19–39

Hallward, Peter (1997) 'Deleuze and the "World Without Others"', *Philosophy Today* (Winter): 530–44

Haraway, Donna (1991) 'A Cyborg Manifesto', from *Simians, Cyborgs, and Women: The Reinvention of Nature* (London: Free Association Books), pp. 149–81

Hardt, Michael (1996) 'Introduction: Laboratory Italy', in P. Virno and M. Hardt (eds.) *Radical Thought in Italy: A Potential Politics* (Minneapolis: University of Minnesota Press)

Hardt, Michael and Antonio Negri (1994) *Labor of Dionysus: Critique of the State Form* (Minneapolis: University of Minnesota Press)

—— (2000) *Empire* (Cambridge, Mass.: Harvard University Press)

—— (2001) 'Response', *Rethinking Marxism*, 13(3/4): 236–43

—— (2002) 'Michael Hardt and Antonio Negri Interviewed by Nicholas Brown and Imre Szeman', *Cultural Studies*, 16(2): 177–92

Hartsock, Nancy (1990) 'Foucault on Power: A Theory for Women?', in L. Nicholson (ed.) *Feminism/Postmodernism* (New York and London: Routledge), pp. 157–75

Hegel, G.W.F. (1977/1807) *Phenomenology of Spirit* (Oxford: Oxford University Press)

Hekma, Gert, Harry Oosterhuis and James Steakley (eds.) (1995) *Gay Men and the Sexual History of the Political Left* (Binghamton, NY: Harrington Park Press)

Hekman, Susan J. (1990) *Gender and Knowledge: Elements of a Postmodern Feminism* (Boston: Northeastern University Press)

Held, David and McGrew, Anthony (2002) *Globalization and Anti-Globalization* (Oxford: Polity Press)

Hennessy, R. (1996) 'Queer Theory, Left Politics', in *Marxism Beyond Marxism* (New York: Routledge), pp. 322–42

Hensman, R. (1996) 'The Role of Women in the Resistance to Political Authoritarianism in Latin America and South Asia', in Haleh Afshar (ed.) *Women and Politics in the Third World* (London: Routledge)

Hewitt-White, Caitlin (2001) 'Women Talking about Sexism and Oppression in the Anti-Globalization Movement', *Kick it Over*, 39

Hobbes, Thomas (1996/1651) *Leviathan* (Cambridge: Cambridge University Press)

Hodgen, M. (1964) *Early Anthropology in the 16th and 17th Centuries* (Philadelphia: University of Pennsylvania Press)

Holloway, John (2002) *Change the World without Taking Power* (London: Pluto Press)

Honneth, Axel (1995) *The Struggle for Recognition* (Cambridge: Polity Press)

hooks, bell (1984) *Feminist Theory from Margin to Center* (Boston: South End Press)

Huff-Hannon, Joseph (2003) 'East Coast Autonomista Caravan'. Available from: <http://www.zmag.org/content/print_article.cfm?itemID=4584&sectionID=1> [accessed July 7, 2004]

IMC Encuentro Proposal Working Group (2000) *Toward and IMC Encuentro*. Available from: <http://lists.indymedia.org/imc-announce> [accessed July 7, 2004]

Jardine, Alice (1985) *Gynesis: Configurations of Women and Modernity* (Ithaca: Cornell University Press)

Johansen, B. (1999) *Native America and the Evolution of Democracy* (London: Greenwood Press)

Jordan, John (1997) 'The Street Party is only a Beginning ... Interview with John Jordan by Naomi Klein', in *The Anarchives*, 4(9). Available from: <www.ainfos.ca/A-Infos97/4/0552.html> [accessed July 7, 2004]

Jordan, Tim (2002) *Activism! Direct Action, Hacktivism and the Future of Society* (London: Reaktion Books)

Kaufman, Vincent (1997) 'Angels of Purity', *October* 79 (Winter): 49–68

Kellner, Douglas (1990) *Television and the Crisis of Democracy* (Boulder and San Francisco: Westview Press)

Kingsworth, Paul (1999) 'India Cheers while Monsanto Burns', *The Ecologist*, 29(1)

Klein, Hilary (2001) 'Women and Indigenous Autonomy', *Bulletin 'Chiapas Al Dia' #242*. Available from: <http://www.ciepac.org/bulletins/ingles/ing242.htm> [accessed July 16, 2004]

Klein, Naomi (2001) 'Reclaiming the Commons', *New Left Review*, 9: 81–9

Knabb, Ken (ed.) (1981) 'Definitions', *Situationist International Anthology* (Berkeley: Bureau of Public Secrets)

Kornegger, Peggy (2002) 'Anarchism: The Feminist Connection', in Dark Star (ed.) *Quiet Rumours: An Anarcha-Feminist Reader* (London: AK Press)

Kropotkin, Petr (1890) *Anarchist Morality* (London: Freedom Press)

—— (1912) *Modern Science and Anarchism* (London: Freedom Press)

—— (1942/1899) *Memoirs of a Revolutionist*, in H. Read (ed.) *Kropotkin: Selections From His Writings* (London: Freedom Press)

—— (1968/1924) *Ethics: Origin and Development* (New York and London: Benjamin Blom)

—— (1989/1902) *Mutual Aid* (Montreal: Black Rose Books)

—— (1990/1892) *The Conquest of Bread* (Montreal: Black Rose Books)

Kubrin, David (2001) 'Scaling the Heights to Seattle', in E. Yuen et al. (eds.) *The Battle of Seattle: The New Challenge to Capitalist Globalization* (New York: Soft Skull Press)

Kymlicka, Will (1995) *Multicultural Citizenship* (Oxford: Clarendon Press, 1995)

—— (1998) *Finding Our Way: Rethinking Ethnocultural Relations in Canada* (Toronto: Oxford University Press)

Lacan, Jacques (1992) *The Ethics of Psychoanalysis* (London: Routledge)

Laclau, E. (1996) *Emancipation(s)* (London: Verso)

Laclau, E. and Mouffe, C. (1985) *Hegemony and Socialist Strategy: Towards a Radical Democratic Politics*, (London: Verso)

Landauer, Gustav (1978/1911) *For Socialism* (St. Louis: Telos Press)

Landless People's Movement (LPM) (2001) 'What the Landless Say in 2001: LPM Founding Statement and Resolutions'. Available from: <http://www.nlc.co.za/pamphlets/lr11.htm> [accessed August 18, 2004]

—— (2004) 'NO LAND! NO VOTE!' Press statement issued April 13, 2004. Available from: < http://lists.iww.org/pipermail/iww-news/2004-April/004798.html> [accessed August 18, 2004]

LaSelva, Samuel and Richard Vernon (1998) 'Liberty, Equality, Fraternity ... and Federalism', in Martin Westmacott and Hugh Mellon (eds.) *Challenges to Canadian Federalism* (Scarborough: Prentice Hall)

Lazzarato, Maurizio (1996) 'Immaterial Labor', in P. Virno and M. Hardt (eds.) *Radical Thought in Italy: A Potential Politics* (Minneapolis: University of Minnesota Press)

Lefort, C. (1988) *Democracy and Political Theory* (Minneapolis: University of Minnesota Press)

Leiss, William (1972) *The Domination of Nature* (Boston: Beacon Press)

Lenin, V. I. (1955/1919a) 'On Democracy and Dictatorship', in *To the Population / On Democracy and Dictatorship / What is Soviet Power?* (Moscow: Foreign Languages Publishing House)

—— (1955/1919b) 'Report on the Right of Recall', in *To the Population / On Democracy and Dictatorship / What is Soviet Power?* (Moscow: Foreign Languages Publishing House)

—— (1963/1911) 'Reformism in the Russian Social-Democratic Movement', *Collected Works*, Vol. 17 (Foreign Languages Publishing House: Moscow), pp. 229–40

—— (1966/1921) 'Fourth Anniversary of the October Revolution', *Collected Works* Vol. 33 (Foreign Languages Press: Moscow), pp. 51–9

—— (1966/1923) 'Better Fewer, But Better', *Collected Works*, Vol. 33 (Moscow: Progress Publishers), pp. 487–502

—— (1967/1902) *What is to be Done?* (Moscow: Progress Publishers)

—— (1975 /1905) *Two Tactics of Social Democracy in the Democratic Revolution* (Peking: Foreign Language Press)

Leonard, Autumn, Tomás Aguilar, Mike Prokosch and Dara Silverman (eds.) (2001) ' Local and Global Organizing after 9/11' Available from: <http://comm-org.utoledo.edu/papers2001/localglobal.htm> [Accessed July 17, 2004]

Leoncavallo Occupation Committee (1975) *18 ottobre 1975—L'occupazione: il primo volantino*. Available from: <http://www.ecn.org/leoncavallo/storic/primvol.htm> [Accessed July 17, 2004]

Locke, John (1988/1690) *Two Treatises on Government* (Cambridge: Cambridge University Press)

London Action Resource Centre (2003) *Social Centre Network* Available from: <http://www.londonarc.org/social_centre_network.html> [accessed October 28, 2003]

Long, J. Anthony, Leroy Little Bear, and Menno Boldt (1984) 'Federal Indian Policy and Indian Self-Government in Canada', in Little Bear et al. (eds.) *Pathways to Self-Determination: Canadian Indians and the Canadian State* (Toronto: University of Toronto Press)

Lorde, Audre (1984) *Sister Outsider* (New York: The Crossing Press)

los Ricos, Rob (2002) 'Empire for Beginners', in *Anarchy: A Journal of Desire Armed* (Spring/Summer)

Lotringer, Sylvere and Marazzi, Christian (1980) 'The Return of Politics', *Semiotext(e) #3 Italy: Autonomia*

Luby, Shawn (2001) in Leonard, Autumn, Tomás Aguilar, Mike Prokosch and Dara Silverman (eds.) (2001) 'Local and Global Organizing after 9/11'. Available from: <http://comm-org.utoledo.edu/papers2001/localglobal.htm> [accessed July 17, 2004]

Lukács, G. (1971) *History and Class Consciousness* (London: Merlin)

MacKenzie, I. (1997) 'Creativity as Criticism: The Philosophical Constructivism of Deleuze and Guattari', *Radical Philosophy*, 86 (Nov.–Dec.): 7–18

MacKinnon, Catharine A. (1989) *Toward a Feminist Theory of the State* (Cambridge, Mass.: Harvard University Press)

Manuel, Frank (1971) Foreword to *Design for Utopia: Selected Writings of Charles Fourier* (New York: Schocken Books)

Maracle, Lee (1996) *I am Woman: A Native Perspective on Sociology and Feminism* (Vancouver: Press Gang Publishers)

Marazzi, Christian (2002) 'Intervista a Christian Marazzi', [CD-ROM] (2000), from the accompanying Borio, Guido, Francesca Pozzi and Guido Roggero, *Futuro Anteriore—Dai 'Quaderni Rossi' ai Movimenti Globali: Ricchezze e Limiti dell'Operaismo Italiano* (Rome: Derive Approdi)

Marco (2000) 'What Moves Us?' *A-Infos Archives*. Available from: <http://www.ainfos.ca/00/aug/ainfos00252.html> [accessed August 21, 2000]

Marcos, Subcomandante Insurgente (2003) *I Shit on All the Revolutionary Vanguards of this Planet*. Available from: <http://flag.blackened.net/revolt/mexico/ezln/2003/marcos/etaJAN.html> [accessed November 3, 2003]

Marcuse, Herbert (1964) *One-Dimensional Man* (Boston, Mass.: Beacon Press)

Marcuse, Peter (2000) 'The Language of Globalization', *Monthly Review*, 52(3): 23–7

Marshall, Peter H. (1992) *Demanding the Impossible: A History of Anarchism* (London: HarperCollins)

Marshall, Stephen (1999) 'Christiania: a Fascinating Place, but Would You Want One on Your Doorstep?', *The Planning Factory* #10 (Summer). Available from: <http://www.bartlett.ucl.ac.uk/planning/information/PF/PF10.html> [accessed August 12, 2004]

Martinez, Elizabeth (2002) 'Where was the Color in Seattle?', in M. Prokosch and L. Raymond (eds.) *The Global Activist's Manual* (New York: Thunder's Mouth Press/Nation Books)

Marule, Marie Smallface. (1984) 'Traditional Indian Government: Of the People, by the People, for the People, in Little Bear et al. (eds.) *Pathways* to Self-Determination: Canadian Indians and the Canadian State (Toronto: University of Toronto Press)

Marx, Karl (1978/1874–75) 'After the Revolution: Marx Debates Bakunin', in Robert Tucker (ed.) The Marx–Engels Reader (New York: Norton)

—— (1975/1847) The Poverty of Philosophy (Moscow: Progress Publishers)

Marx, Karl and Engels, Friedrich (1978/1848) 'The German Ideology', in Robert Tucker (ed.) The Marx–Engels Reader (New York: Norton)

—— (1888/1848) Manifesto of the Communist Party (Moscow: Progress Publishers)

Matisz, Paul (2003) Interview with Sean Haberle, for the Affinity Project, Queen's University at Kingston. Conducted June 17, 2003 at Peterborough, Ontario

May, T. (1994) The Political Philosophy of Poststructuralist Anarchism (University Park, Pennsylvania: University of Pennsylvania Press)

Melucci, Alberto (1989) Nomads of the Present. London: Hutchinson Radius

Millàn, Márrgara (1998) 'Zapatista Indigenous Women', in J. Holloway and E. Peláez (eds.) Zapatista: Reinventing Revolution in Mexico (London: Pluto Press)

Milstein, Cindy (2002) 'What's in a Name?', Harbinger, 2(1). Available from: <http://www.social-ecology.org/harbinger/vol2no1/name.html> [accessed March 18, 2002]

Minh-Ha, Trinh T. (1991) When the Moon Waxes Red: Representation, Gender and Cultural Politics (New York: Routledge)

Mishra, Pramod K. (2001) 'The Fall of the Empire or the Rise of the Global South?', Rethinking Marxism, 13(4): 100–18

Mitchell, Grand Chief Michael (1989) 'Akwesasne: An Unbroken Assertion of Sovereignty', in Boyce Richardson (ed.) Drum Beat: Anger and Renewal in Indian Country (Toronto: Summerhill Press), pp. 109–10

Mohanty, Chandra (2003) Feminism without Borders: Decolonizing Theory, Practicing Solidarity (Durham, NC and London: Duke University Press)

Monsanto, Inc. (2002) 'Mahyco-Monsanto Plans Big Push for Bt Cotton', Agbioworld, Friday April 19, 2002

Monture-Angus, Patricia (1999) Journeying Forward: Dreaming First Nations' Independence (Halifax: Fernwood)

Mookerjea, Sourayan (2003) 'Migrant Multitudes, Western Transcendence and the Politics of Creativity', Journal for Cultural Research 7(4): 405–32

Moore, David (2001) 'Africa: The Black Hole at the Middle of Empire?', Rethinking Marxism, 13(4): 100–18

Moraga, C. and G. Anzaldúa (eds.) (1981, 2nd edn. 1983) This Bridge Called My Back: Writings by Radical Women of Color (New York: Kitchen Table: Women of Color Press)

Morton, D. (2001) 'Global (Sexual) Politics, Class Struggle, and the Queer Left', in Post-Colonial, Queer: Theoretical Intersections (Albany, NY: SUNY Press)

Mouffe, C. (1993) The Return of the Political (London: Verso)

MST (Landless Worker's Movement) (2003) 'Homepage'. Available from: <http://www.mstbrazil.org/index.html> [Accessed September 2, 2003]

Mutume, Gumisai (2000) Ghanaians Opposing Water Privatization. Available from: <http://www.nadir.org/nadir/initiativ/agp/campanas/water/txt/2000/1115Ghana.htm> [accessed September 2, 2003]

Newman, Saul (2001) *From Bakunin to Lacan: Anti-Authoritarianism and the Dislocation of Power* (Lanham: Lexington Books)

Nicholson, Linda (ed.) (1989) *Feminism/Postmodernism* (London: Routledge)

NOII (No One is Illegal) Montreal (2004) *Justice & Dignity For (Im)migrants, Refugees & Indigenous Peoples* (Montreal: NOII)

OCAP (Ontario Coalition Against Poverty) (2002) (ocap@tao.ca) email from Ontario Coalition Against Poverty (ocap@lists.tao.ca)

Offe, C. (1985) 'New Social Movements: Challenging the Boundaries of Institutional Politics', *Social Research*, 52(4): 817–68

Owen, Robert (1973/1849) *The Revolution in the Mind and Practice of the Human Race* (Clifton, NJ: Augustus M. Kelley)

Pablo (2004) 'Argentines Speak Out: Voices from the Neighbourhood Assemblies'. Available from: <http://argentinanow.tripod.com.ar/6.html> [accessed July 3, 2004]

Pachamama Alliance (2003) 'Relations Strain Between CONAIE and Ecuador's President—Oil and Foreign Debt Among Issues Negotiated', in *New Moon Update*. Available from: <http://www.pachamama.org/updates/new-moon-2003-jul.htm> [accessed September 2, 2003]

Panitch, Leo (2001) *Renewing Socialism: Democracy, Strategy, and Imagination* (Boulder: Westview Press)

Paredes, Julieta (2002) 'An Interview with Mujeres Creando', in Dark Star (ed.) *Quiet Rumours: An Anarcha-Feminist Reader* (London: AK Press)

Parekh, Bhikhu (2000) *Rethinking Multiculturalism: Cultural Diversity and Political Theory* (London: Macmillan Press)

Pastor, Manuel and LoPresti, Tony (2004) 'Bringing Globalization Home: Lessons from Miami in Projecting the Voices of People of Colour and Connecting Global Forces to Local Problems', in *ColorLines* (Summer) pp. 29–32

Pearson, K. A. (1998) 'Freedom and the Politics of Desire: Aporias, Paradoxes, and Excesses', *Political Theory* 26(3): 399–412

People's Global Action (PGA) (2002) 'We are Everywhere—Even Aotearoa!', *thr@ll* #23. Available from: <http://free.freespeech.org/thrall/23pga1.html> [accessed May 23, 2002]

—— (2003) 'IMF, World Bank, and WTO: Policies and Resistance'. Available from: <http://www.nadir.org/nadir/initiativ/agp/free/imf/> [accessed February 3, 2003]

—— (2004a) *Hallmarks of People's Global Action*. Available from: <http://www.nadir.org/nadir/initiativ/agp/free/pga/hallm.htm> [accessed February 9, 2004]

—— (2004b) Proceedings of Third International Conference of Peoples. Global Action (PGA), Cochabamba, Bolivia, September 16–23, 2001, Appendix 4: Organizational Principles of the Peoples' Global Action (PGA). Available from: <http://www.nadir.org/nadir/initiativ/agp/cocha/cocha.htm#ap4> [accessed July 17, 2004]

Perez, Rolando (1990) *On An(archy) and Schizoanalysis* (New York: Semiotext(e))

Petras, James (2000) 'The Rural Landless Worker's Movement'. Available from: <http://www.zmag.org/Zmag/articles/march2000petras.htm> [accessed May 23, 2002]

Povinelli, Elizabeth (1998) 'The State of Shame: Australian Multiculturalism and the Crisis of Indigenous Citizenship (Intimacy)', *Critical Inquiry*, 24(2): 575–611

Press, Andrea Lee (1991) *Women Watching Television: Gender, Class and Generation in the American Television Experience.* (Philadelphia: University of Pennsylvania Press)

Project Censored (2004) *#23 Argentina Crisis Sparks Co-operative Growth.* Available from: <http://www.projectcensored.org/publications/2004/23.html> [accessed July 15, 2004]

Proudhon, P. J. (1923/1851) *General Idea of the Revolution in the Nineteenth Century* (London: Freedom Press)

—— (1969) *Selected Writings of Pierre Joseph Proudhon*, ed. Stewart Edwards (New York: Anchor Books)

—— (1971/1863) *The Principle of Federation* (Toronto: University of Toronto Press)

Pulido, Laura (1998) *Environmentalism and Economic Justice: Two Chicano Struggles in the Southwest* (Tucson: University of Arizona Press)

Pybus, Jennifer (2003) 'Emilia Romagna Cooperatives: Political-Economic and Cultural Alternatives within a Capitalist Paradigm'. Unpublished Honour's Thesis, Dept. of Communication, Simon Fraser University, presented December 2002

Quinby, Lee (2003) 'Taking the Millennialist Pulse of Empire's Multitude: A Genealogical Feminist Analysis', in Paul Passavant and Jodi Dean (eds.) *Empire's New Clothes: Reading Hardt and Negri* (London: Routledge)

Rebick, Judy (2002) 'Lip Service: The Anti-Globalization Movement on Gender Politics', *Herizons*, 16(2) pp. 24–6

Reclaim the Streets (RTS) London (2000a) 'On Disorganization: A Statement from Reclaim the Streets (RTS) London'. Released Friday April 14, 2000. Available from: <http://www.reclaimthestreets.net/> [accessed June 13, 2002]

—— (2000b) *Street Politics: A Response to the Mayday 2000 Media.* Available from: <http://rts.gn.apc.org/streetpolitics.htm> [Accessed June 13, 2002]

Rissanovsky, Nicholas V. (1969) *The Teachings of Charles Fourier* (Berkeley: University of California Press)

Said, Edward (1978) *Orientalism* (New York: Vintage Books)

Saint-Simon, Claude Henri comte de (1976/1823–26) *The Political Thought of Saint-Simon*, ed. G. Ionescu (London: Oxford University Press)

—— (1975/1825) 'The New Christianity', in K. Taylor (ed. and trans.) *Henri Saint-Simon: Selected Writings on Science, Industry, and Social Organization* (New York: Holmes and Meier)

Salvadori, M. (1979) 'Gramsci and the PCI: Two Conceptions of Hegemony', in C. Mouffe (ed.) *Gramsci and Marxist Theory* (London: Routledge)

Schechter, D. (1994) *Radical Theories: Paths beyond Marxism and Social Democracy* (Manchester: Manchester University Press)

SCP/NYC (2004a) 'Time in the Shadows of Anonymity: Against Surveillance, Transparency and Globalized Capitalism'. Available from: <http://www.notbored.org/transparent.html> [accessed July 17, 2004]

SCP/NYC (2004b) *Who We Are and Why We're Here*. Available from: <http://www.notbored.org/generic.jpg> [accessed July 17, 2004]

Sedgwick, Eve Kosofsky (1990) *Epistemology of the Closet* (Berkeley and Los Angeles: University of California Press)

Seok, Kyong-Hwa (2000) 'South Koreans Protest Restructuring'. Available from: <http://www.nadir.org/nadir/initiativ/agp/free/imf/asia/skorea.htm#restructuring> [accessed April 24, 2002]

Shalem, Yael and David Bensusan (1994) 'Civil Society: The Traumatic Patient', *Angelaki*, 1(3): 73–92

Shiva, Vandana (1988) *Staying Alive: Women, Ecology and Development in India* (London: Zed Books)

Shore, Chris (1998) 'Inventing the "People's Europe": Critical Approaches to European Community "Cultural Policy"' in *Man*, 28(4)

Singh, Jaggi (2001) 'Resisting Global Capitalism in India', in Eddie Yuen et al. (eds.) *The Battle of Seattle: The New Challenge to Capitalist Globalization* (New York: Soft Skull Press)

Smith, Anna Marie (1998) *Laclau and Mouffe: The Radical Democratic Imaginary* (London: Verso)

Smith, Dorothy (1977) *Feminism and Marxism: A Place to Begin, a Way to Go* (Vancouver: New Star Books)

Smith, Sharon (1994) 'Mistaken Identity—or Can Identity Politics Liberate the Oppressed?', *International Socialism Journal*, 62 (Spring)

Spivak, Gayatri (1988) 'Can the Subaltern Speak?' in C. Nelson and L. Grossberg (eds.) *Marxism and the Interpretation of Culture* (Urbana: University of Illinois Press), pp. 271–313

Spurlin, William J. (2001) 'Broadening Postcolonial Studies/Decolonizing Queer Studies: Emerging 'Queer' Identities and Cultures in Southern Africa', in *Post-Colonial, Queer: Theoretical Intersections* (Albany, NY: SUNY Press)

Stavrakakis, Yannis (1999) *Lacan and the Political* (London: Routledge)

Stedile, Joan Pedro (2002) 'Landless Battalions: The Sem Terra Movement of Brazil', *New Left Review*, 15: 77–104

Stephen, Lynn (1995) 'The Zapatista Army of National Liberation and the National Democratic Convention' *Latin American Perspectives*, 22(4): 88–99

Styles, Sophie (2002) 'An Interview with Julieta Ojeda of Mujeres Creando', *Z Magazine* 15:6. Available from: <http://www.zmag.org/ZMag/articles/jun02styles.html> [accessed July 27, 2004]

Sweezy, Paul M. and Magdoff, Harry (2000a) 'Toward a New Internationalism', in *Monthly Review* 52(3): 1–10

—— (2000b) 'Socialism: A Time to Retreat?' *Monthly Review* 52(4): 1–17

Taris, James (2003) 'LETS Changed My Life', *New Community Quarterly* (May)

Taylor, Charles (1975) *Hegel* (Cambridge: Cambridge University Press)

—— (1984) 'Foucault on Freedom and Truth', *Political Theory*, 12 (March): 152–83

—— (1991) *The Malaise of Modernity* (Concord: Anansi)

—— (1992) 'The Politics of Recognition', in A. Gutman (ed.) *Multiculturalism and the Politics of Recognition* (Princeton: Princeton University Press)

—— (1993) *Reconciling the Solitudes: Essays on Canadian Federalism and Nationalism* (Montreal and Kingston: McGill-Queen's University Press)

—— (1998) 'The Dynamics of Democratic Exclusion', *Journal of Democracy*, 9(4)

Taylor, Keith (1975) *Henri Saint-Simon (1760–1825): Selected Writings on Science, Industry, and Technology* (New York: Holmes and Meier)

Thomassen, Benjamin S. (2002) 'Rave = TAZ'. Available from: <http://www. chikinsandowichi.com/ravedata/taz5.htm> [accessed April 28, 2002]

Touraine, Alain (1992) 'Beyond Social Movements?', *Theory, Culture and Society* 9: 125–45

Trask, Haunani-Kay (1993) 'Environmental Racism in Hawai'i and the Pacific Basin'. Talk delivered at the University of Colorado, Boulder, September 29, 1993. Available from: <http://www.zmag.org/ZMag/articles/bartrask. htm> [accessed April 28, 2002]

Unanimous Consensus (2001) 'Organizing a New Generation of Activists: The Nuts and Bolts of Working Together', in Neva Welton and Linda Wolf (eds.) *Global Uprising: Confronting the Tyrannies of the 21st Century* (Gabriola Island: New Society Publishers)

Unemployed Workers of Lanus et al. (2003) *Argentina: The 'Piqueteros' of the Unemployed Workers Movement.* Available from: <http://www. solidaridadesrebeldes.kolgados.com.ar/article.php3?id_article=24> [Accessed July 10, 2004]

urban75 (2003) *Smashing the Image Factory: A Complete Manual of Billboard Subversion & Destruction.* Available from: <http://www.urban75.com/Action/ factory.html> [Accessed August 19, 2003]

Uzelman, Scott (2002) 'Catalyzing Participatory Communication: Independent Media Centre And The Politics of Direct Action'. MA Thesis, Burnaby, BC: Simon Fraser University

Virno, Paolo (1996) 'Virtuosity and Revolution: The Political Theory of Exodus', in P. Virno and M. Hardt (eds.) *Radical Thought in Italy: A Potential Politics* (Minneapolis: University of Minnesota Press)

Ward, Colin (1996) *Anarchy in Action* (London: Freedom Press)

Werbe, Peter (1999) Interview with Keith McHenry in 'Midwestern Anarchists Speak Out!', *Alternative Press Review* 3(3) (Spring/Summer)

Wetzel, Tom (n.d.) 'Participatory Economics and the Self-Emancipation of the Working Class', *The Participatory Economics Project.* Available from: <http://www.parecon.org/writings/wetzel_emancipation.htm> [accessed November 17, 2003]

Wheeler, Anna and Thompson, William (1970/1825) *Appeal of One Half of the Human Race, Women, against the Pretensions of the other Half, Men …* (New York: B. Franklin)

Wright, Steve (2002) *Storming Heaven: Class Composition and Struggle in Italian Autonomist Marxism* (London: Pluto Press)

Young, I. M. (1989) 'Polity and Group Difference: A Critique of the Ideal of Universal Citizenship', *Ethics*, 99

Zerzan, John (2002) 'Against Technology', in *Running on Emptiness* (Los Angeles: Feral House)

—— (n.d.) 'The Catastrophe of Postmodernism'. Available from: <http://www.
primitivism.com/postdmoernsism.htm> [accessed November 17, 2003]

Ziarek, Eva P. (2001) *An Ethics of Dissensus: Postmodernity, Feminism, and the
Politics of Radical Democracy* (Stanford: Stanford University Press)

Žižek, S. (1991) *Looking Awry: An Introduction to Jacques Lacan through Popular
Culture* (Cambridge, Mass.: MIT Press)

—— (1994) *The Metastases of Enjoyment* (London: Verso)

—— (1997) 'Multiculturalism, or, the Cultural Logic of Late Capitalism', *New
Left Review*, 225: 28–51

Index